# FAQS in STATS

## Frequently Asked Questions in Introductory Statistics

### Examples Include the TI-84, R, and Excel

## Isabella Romeo

Sherwood
Forest

http://CalculusCastle.com

FAQS in STATS: Frequently Asked Questions in Introductory Statistics
by Isabella Romeo
First Edition
Last Revised on: 5/1/17 (Version 1.0)


Sherwood Forest Books
Los Angeles, CA, USA


ISBN-13: 9781540546036
ISBN-10: 1540546039

Statistics books suck. Don't expect this one to be any better. Please report any errors, omissions, or suggestions for improvements through the form at https://github.com/bellaromeo/faqsinstats/issues/new. Please include the version number and revision date listed at the top of this page on all communications.

# Table of Contents

# Comparing Variables   201

# Describing Relationships   231

# Some Final Thoughts   253

# Stuff in the Back of the Book                                       264

**About software licensing.** No software is required to accompany this book. Software is used in the technology examples, but these may be skipped without loss of continuity. Should you choose to use any of the software referred to in this book, please read the software license carefully before you decide to install it on your computer.

R as a package is free software that is licensed under the Gnu Public License (GPL); other licenses apply to parts of R and some associated software. Details are given at https://www.r-project.org/Licenses/. It is available for Windows, Mac and Linux.

RStudio is free software that is licensed under the GNU Affero General Public License (AGPL v3), as described at https://support.rstudio.com/. RStudio is an open source product. Commercial versions are also available. It is available for Windows, Mac and Linux.

LibreOffice is available under the Mozilla Public License Version 2.0. See https://www.libreoffice.org/about-us/licenses/. It is open source and free to download and share. It is available for Windows, Mac and Linux.

OpenOffice (version 4) is available under the Apache License 4.0 as described at https://www.openoffice.org/license.html. Earlier versions (version 3) were licensed under the Lesser Gnu Public License (LGPL) version 3. It is open source and free to download and share. It is available for Windows, Mac and Linux.

Microsoft Excel is a commercial product and you must purchase a license in order to use it on your computer. You may not share copies with other people and cannot freely download it over the internet. The details of the license vary, depending on how you purchased the product. In general, you do not even own the copy of the software that you purchased, only a license to use it, sometimes only for some limited period of time. For details, see https://www.microsoft.com/en-us/useterms. It is only available on Windows and Mac operating systems.

# Preface

This book is designed for students, to accompany a typical first year (e.g., college Freshman) course in statistics. The only mathematical background expected from students is some understanding of basic high school algebra. It can be used as a workbook in a lab course or as a study guide for students. It is not designed to be a primary textbook.

**Organization.** Each chapter is organized around a single **question**, such as "How do I find the variance?" Sometimes a question has multiple parts, as in "What is a density function? What is the difference between a density and a distribution?" The chapter attempts to answer the question, and where appropriate, provides a **step-by-step procedure** for computing the object in question. The computation can always be computed using a basic calculator along with tables of known statistical functions (such as the cumulative density function of the standard normal distribution). The tables are provided in the **appendices**. A single **example** is always integrated within the procedure; wherever the procedure bifurcates (takes different possible directions), additional examples are provided. Where appropriate, the procedure is then followed by **technology examples**. The chapter ends with a collection of short **exercises**.

**Technology.** The purpose of the technology examples is to give students a quick and dirty way of getting statistical results using technology. These examples answers questions such as: How do I do this in R? How do I do this in a spreadsheet? How do I do this on a scientific calculator? In general, these examples can (and should) be totally ignored unless you already have access to and are required to continue using one of these three forms of technology in your stats class.

▸ Using the **TI-84 calculator**. In general, I discourage students from spending money on expensive calculators like the TI-84 unless they are engineering majors. If your teacher lets you use this type of calculator, it will make your work considerably easier. However, if you are taking your first and last stats course, you will never, ever use the calculator again. Nobody uses calculators professionally.[1]

- Using the **R-language**. This is how the vast majority of professionals do statistics. R is a free and very easy to learn computer language. The number of keystrokes you have to type in for each calculation is comparable to the number of keystrokes on the TI-84. If you want to get dangerous with statistics, you need to learn some R. An easy to use implementation is RStudio.[2]

- Using **Excel (or equivalent free[3]) spreadsheets**. Virtually every employer will expect you to know how to use Excel (or an equivalent free) spreadsheet. It is not as powerful as R, but once your know Excel, there are ways to link it to R (and vice-versa) so that you can include whatever you need.[4]

The technology examples are simply precise recipes for solving the same problems that are also solved by hand. They can be generalized by the substitution of other numbers but do not provide overall instructions on how to use the software or technology for other purposes.

**Book Version.** Thank you for buying the book. If you are reading an electronic version of this book but did not purchase it at Gumroad (http://www.gumroad.com), then you are reading an illegal copy. Please go to Gumroad and purchase a DRM-free copy.

**Feedback.** I welcome your comments and suggestion. Please let me know what you think about this book, any errors you find, and any suggestions you might have for improvement. You can reach me by raising an issue at the Github site associated with this book. There is a form for this at https://github.com/bellaromeo/faqsinstats/issues/new. If you don't have a Github account you can get one for free (no strings attached).

---

[1] The TI-84 examples were all developed using a TI-84-Plus-CE, so if you have a regular TI-84 or a different model TI-calculator, the syntax might be different.

[2] See https://www.rstudio.com/.

[3] They all work the same way. If you don't want to pander to the evil empire, you can get free versions from LibreOffice (https://www.libreoffice.org/) or Open Office (https://www.openoffice.org/) that are completely compatible with Excel. Even Apple has its own (non-free) software (Numbers, http://www.apple.com/numbers/) but it is not widely used.

[4] Not covered in this book.

This book is dedicated to the memory of Bella's life partner Romeo, who succumbed to a sudden and unexpected infection on 24 April 2017, just as the text was completed. Always the loyal friend, he hung on until every chapter was typeset, not telling anyone whether or not he had a tummy-ache, just giving love and comfort to those around him, up until the very end. I imagine him now in the Summerlands, lounging in the sun and chasing squirrels all day long.

B.E.S.

# FAQ 1. What is Statistics?

Statistics is the study of data. It can be a very powerful tool. When used incorrectly it can be a dangerous tool, because how you manipulate the data may affect how you interpret it. It is often said that there are three kinds of lies: lies, damn lies, and statistics.[1] Psychoanalyst Wilhelm Stekel (1868-1940) put it this way: "statistics is the art of lying by means of figures." Statistician Carroll Wright (1840-1909) said "figures don't lie, but liars will figure."

The interpretation of statistical results is rarely clear cut. For example, a common joke goes like this: "Four out of five accidents occur at home. So why do so many people still live there?" Another joke refers to the famous American President Abraham Lincoln, who was once heard saying that three quarters of the information posted on the internet is bogus. As author Rex Stout (1886-1975) would have us believe, "There are two kinds of statistics; the kind you look up and the kind you make up" (*Death of a Doxy*).

This book focuses on two areas of statistics that are commonly studied in introductory courses:

- **Descriptive statistics**: describing the distribution and properties of collection of data.

  Examples include finding means, standard deviations, and medians; making plots; describing associations; correlations; slopes and intercepts.

- **Inferential statistics**: predicting a property of a population from a measurement of a random sample.
  Examples include confidence intervals and hypothesis tests.

*FAQs in Stats* is not intended to be a primary textbook in statistics. Presumably you are studying statistics and already have another textbook that explains things in suitably excruciating detail. *FAQs in Stats*

---

[1] Often attributed to both Mark Twain and Benjamin Disraeli. In fact, the original quote refers to three kinds of witnesses: liars, damned liars, and experts, and was published in an editorial in *Nature* in Nov. 1885.

is more of a workbook that can accompany your textbook, giving you procedures for calculation.

Furthermore, *FAQs in Stats* is a book of frequently asked questions about statistics. Each chapter title is actually a question that has been posed during a class session. The overall organization of the book follows the same order that a typical statistics text might follow. While many of the sections are written so that they can be read stand-alone, learning statistics is a cumulative process. For example, to calculate the slope of a line of least squares requires the correlation; the correlation requires the calculation of two different standard deviations; and each standard deviation requires the calculation of a mean. So don't expect to jump right into the middle and be able to understand everything at once. Good reading!

## FAQ 2. What are Individuals and Variables?

When we collect data, the information we collect represents something. One way we can organize information is in a table, analogous to a spreadsheet.

**Example 2.1.** Suppose a college database contains student information such as name, class level, grade point average, age, and gender. Some of this information might look like this:

| Name | Class | GPA | Age | Gender |
|------|-------|-----|-----|--------|
| Marlene Adams | Freshman | 3.6 | 19 | F |
| Charles Brooks | Sophomore | 2.9 | 22 | M |
| April Gonzales | Senior | 3.2 | 25 | F |
| Sam Cook | Sophomore | 1.8 | 19 | M |
| Destiny Sarceno | Senior | 3.9 | 22 | F |
| Abe Thomas | Junior | 2.3 | 29 | M |

Information in data is classified in two ways. These might be thought of as the dimensions, or directions, through a data table.

- ▸ **Individuals** - a single object, such as a person, a rock, or plant, described in a data set.
- ▸ **Variables** - a property of an object that is allowed to take on different values for different individuals.

**Example 2.2.** In example 2.1 the individuals are students. Each individual is shown on a separate line. An example is

---

| Charles Brooks | Sophomore | 2.9 | 22 | M |

The variables are shown by the rows. They are name, class, GPA, age, and gender. For example, Marlene Adams age is 19, while Charles Brooks' age is 22.

We will consider two types of variables:

- ▸ **Categorical** - the value of the variable can place the individual into an enumerable list of categories. Examples are gender, eye color, and name.
- ▸ **Quantitative** - the value of the variable is numeric, has some sort of units of measurements, and must make sense in an equation (such as averaging). Examples are weight, height, and age. Examples that are categorical and not quantitative are phone number and zip code.

**Example 2.3.** In example 2.1 name, class, and gender are categorical, and GPA and age are quantitative.

# FAQ 3. What are Populations and Samples?

A **population** is the entire group of individuals (ch. 2) about which we are interested.

When we perform a statistical study we select a (usually small) **sample** of the population. We do our data analysis on the sample. In inferential statistics we analyze the sample to infer information about the popula-

Figure 3.1.: Many samples can be be taken from a single population. Each sample has a sample mean $\bar{x}_i$ and sample standard deviation $s_i$ associated with it, that are probably different than the population mean $\mu$ and population mean $\sigma$.

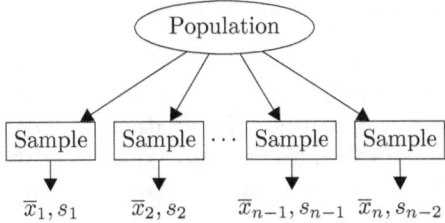

tion. Many samples can be taken from the same population (fig. 3.1).

**Example 3.1.** Consider data of the sort described in 2.1. Suppose you are studying the relationships among class, gender, age, and grade point averages across the entire University of California System. The population would be the entire UC system (approximately 250,000). To make inferences about this entire population the sample would most likely be of several hundred students selected from among the various campuses of the UC system.

Individuals in a sample should be selected randomly. In a **simple random sample (SRS)**

- ▸ Every set of $n$ individuals selected from the same population must be equally likely to be chosen as an SRS.
- ▸ Every individual in the population must be equally likely to be chosen for an SRS.

## FAQ 4. What is the Difference Between a Parameter and a Statistic?

A parameter is a number that describes a population. Often these numbers are given as symbols in the formula or equation for the distribution that describes the population. Typical parameters are the mean $\mu$ and standard deviation $\sigma$ of a population. Another common parameter is a proportion, such as the proportion of women in the United States (the population would be the entire US) or the proportion of the electorate that voted for Richard Nixon in 1968 (the population would be the number of individuals who voted in the presidential election in 1968).

---

**Parameter**

A **parameter** is a number that describes a population.

---

A statistic is a number that is computed from a sample. A sample is only a subset of the population. The statistic is not meaningful unless the sample is random (SRS, chapter 3). Typical statistics are the sample mean $\bar{x}$ (ch. 13), sample standard deviation $s$ (ch. 15), and sample proportion (ch. 49).

Figure 4.1.: Many samples with different statistics might be drawn from the same population. Here the population has mean $\mu$ and standard deviation $\sigma$; the statistics of each sample are $\bar{x}_i$ and $s_i$.

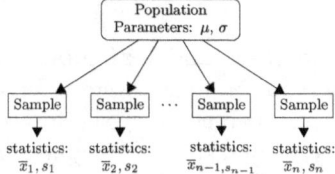

---

## Statistic

A **statistic** is a number that describes a sample.

---

**Example 4.1.** You read that the national average on the Math SAT last year was 514 and standard deviation 93. You take a sample of 25 students in your school to determine how they compare and find that the average of their Math SAT scores is 493, and the standard deviation of their scores is 19.

Which numbers are statistics, which are parameters?

The population is the collection of all students who took the SAT last year, and the sample is the 25 students you interviewed. Since $\mu = 514$ and $\sigma = 93$ describe the population, they are parameters. Similarly, since $\bar{x} = 493$ and $s = 19$ describe the sample, they are statistics.

**Example 4.2.** A news organization conducted exit polls of 742 individuals following the 1968 presidential election. $\boxed{312}$ voters indicated that they had voted for Nixon, $\boxed{311}$ voters indicated that they had voted for Humphrey, and the rest voting for other candidates.

The news agency reported that the the two candidates returns were $\boxed{42.05\%}$ to $\boxed{41.09\%}$, Nixon over Humphrey, with 16% voting for others. All of these boxed numbers are statistics.

In the final tally of 73,199,999 votes, $\boxed{31{,}783{,}783}$ were cast for Nixon, $\boxed{31{,}271{,}839}$ were cast for Humphrey, $\boxed{9{,}901{,}108}$ were cast for Wallace, and $\boxed{243{,}259}$ were cast for other candidates. This gave a slight edge of $\boxed{43.42\%}$ for Nixon over $\boxed{42.72\%}$ for Humphrey. All of the numbers in rounded boxes are parameters, because they described the total population of the electorate.

# FAQ 5.  What is a Stem Plot?

**Stem plots**, also called **stem and leaf plots** use the numbers themselves to visualize the data set. They were invented in the days before computers and are a rarely used nowadays because of the easy availability of software for other types of displays. They are easy to draw by hand, particularly as data is being accrued. Stem plots with more than a few dozen points are extremely unwieldy and either dot plots or histograms should be used instead.

In a stem plot, each data value is separated into a **stem** and a **leaf**. The number of leaves corresponding to each stem will vary, depending on the numbers in the data set. The individual stems and leaves are displayed like this:

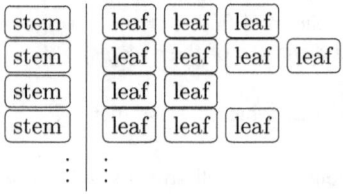

The leaf most frequently consists of the last (rightmost) digit of a number, and the stem the remaining digits. Using this logic

- 96 would have a stem of 9 and a leaf of 6
- 143 would have a stem of 14 and a leaf of 3
- 2487 would have a stem of 248 and and leaf of 7

Sometimes the leafs are rounded, especially if the numbers have many digits and are far apart; an example is given below.

**Example 5.1.** Consider the number 142. We might divide it like this:

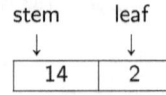

In a stem plot, all of the numbers in the data set with the same stem are collected together in a single place. The stem plot:

```
3 | 49
4 | 36789
5 | 22246
6 |
7 | 158
```

can be read directly as follows: 34, 39 (first row); 43, 46, 47, 48, 49 (second row); 52, 52, 52, 54, 56 (third row); 71, 75, 78 (fifth row). The stem for the first row is 3, and there are two leaves: 4, and 9. These give us 34 and 39. The remaining values are constructed in the same manner.

We demonstrate the process of making stem plots in the following steps.

1. Identify all the stems in the data set. The depth of the leaves depends on the data
   - If the data is entirely 2 digit numbers, use the first digit for the stems and the second for the leaves
   - If the data is entirely 3 digit numbers you have options (skip these examples the first time you read this).
     ▸ Use the first 2 digits as the stems and the third as as the leaves.
       **Example 5.2.** The list (111, 123, 115, 109, 126, 105, 104) could be plotted with stems 10, 11, and 12:

       ```
       10 | 459
       11 | 15
       12 | 36
       ```

     ▸ Use on the first digit as the stem and round the second and third to the nearest ten to use as leaves.
       **Example 5.3.** The list (111, 115, 243, 373, 342, 265, 347, 342) could be plotted with with stems 1,2 and 3. The leaves for 111 and 115 are 1 and 2, respectively, because 111 rounds to 110 and 115 rounds to 120. Similarly 243 has a stem of 2 and leaf of 4, and so forth.

       ```
       1 | 12
       2 | 47
       3 | 4457
       ```

**Example 5.4.** Make a stem plot of the following data set.

78, 48, 52, 71, 47, 56, 52, 54, 39, 49, 52, 34, 43, 75, 46

Since this lists consists entirely of 2-digit numbers, we will use the tens column as stems. The stems in the data are 3, 4, 5 and 7. This is because there are numbers in the 30's, the 40's, the 50's, and the 70's. There are no numbers with stems of 6 because there are no numbers in the 60's.

2. Identify the smallest stem and largest stem, and write them in a vertical column on a sheet of paper. Fill in all possible values in between, and draw a vertical line next to the stems.
   (**Continuation of example 5.4.**) Here are the stems for the data in the example:

   ```
   3 |
   4 |
   5 |
   6 |
   7 |
   ```

3. Add the leafs, one at a time, to the right hand side of the vertical line, adjacent to the appropriate bar.
   (**Continuation of example 5.4.**)   The first number in the list is 78. Its stem is 7, and its leaf is 8:

   ```
   3 |
   4 |
   5 |
   6 |
   7 | 8
   ```

   The second number is 48, with a stem of 4 and a leaf of 8. The third is 52, with a stem of 5, and a leaf of 2.

   ```
   3 |
   4 | 8
   5 | 2
   6 |
   7 | 8
   ```

   The fourth number is 71, with a stem of 7 and a leaf of 1. There is already one leaf in the 7's row so we add a second row. It helps to keep the digits in order, so we put the 1 in front of the 8:

   ```
   3 |
   4 | 8
   5 | 2
   6 |
   7 | 18
   ```

   We continue through the rest of the numbers, adding a 7 to the 4's row; a 6 to the 5's row; a 2 to the 5's 5ow; a 4 to the 5's row; and so on. The final stem plot looks like this:

```
3 | 49
4 | 36789
5 | 22246
6 |
7 | 158
```

There are three 2's in the 5's row because the number 5 appears three times in the data. The 6's row is empty because there are no 6's in the data. Nevertheless, we always leave a blank line when there is missing data.

**Technology Example 5.5 (R).** Make a stem plot of the data in example 5.4 using R.

Use the function **stem**:

```
> data=c(78,48,52,71,47,56,52,54,39,49,52,34,43,
  75,46)
> stem(data)

  The decimal point is 1 digit(s) to the right
  of the |

  3 | 49
  4 | 36789
  5 | 22246
  6 |
  7 | 158
```

Notes:

1. If you data contains decimals and the data points are close together, your stem might be the one's or one-tenth's column.
   **Example 5.6.** If your data looks like:

   6.34, 6.58, 6.48, 6.4, 6.53, 6.43, 6.37, 6.52, 6.43, 6.66, ...

   You might consider using stems of 6.3, 6.4, ...

2. If the data contains larger numbers you might want to round your leafs.
   **Example 5.7.** Consider the following data set:

   4411, 4618, 4781, 4784, 4574, 4515, 4464, 4702, ...

   This might be drawn with stems of 44, 45, 46, ... The leaf 11 in 4411 would be 1; the leaf 18 in 4618 could be a 2 (18 rounds to 20 which truncates to a single digit of 2); the leaf 81 in 4781 becomes an 8; and so forth.

---

3. Sometimes you want to go into more detail, and split each stem at the half-way point.

**Example 5.8.** Construct two different stem plots of the following data:

$$41, 58, 45, 45, 53, 44, 49, 44, 57, 56, 45, 47, 44, 56, 45, 51, 48, 58, 52, 50$$

This be represented either with split stems as

```
4  2
   5 5 6 6 7 8 9
5  0 1 2 3 3 3 4
   8 8
6  2 2 4
```

or with full stems as

```
4  2 5 5 6 6 7 8 9
5  0 1 2 3 3 3 4 8 8
6  2 2 4
```

# Exercises

Make stem plots of each of the following data sets.

1. 78, 48, 52, 71, 47, 56, 52, 54, 39, 49, 52, 34, 43, 75, 46
   Ans:
   ```
   3  4 9
   4  3 6 7 8 9
   5  2 2 2 4 6
   6
   7  1 5 8
   ```

2. 35, 45, 53, 57, 66, 61, 49, 44, 44, 45, 47, 55
   Ans:
   ```
   3  5
   4  4 4 5 5 7 9
   5  3 5 7
   6  1 6
   ```

3. 29, 52, 48, 66, 51, 44, 62
   Ans:
   ```
   2  9
   3
   4  4 8
   5  1 2
   6  2 6
   ```

4. 519, 523, 527, 505, 515, 523, 527, 536
   Ans:
   ```
   50  5
   51  5 9
   52  3 3 7 7
   53  6
   ```

5. 83, 56, 42, 56, 99, 118, 147, 102, 112, 108, 52, 110, 86, 50, 54
   Ans:
   ```
   4   2
   5   0 2 4 6 6
   6
   7
   8   3 6
   9   9
   10  2 8
   11  0 2 8
   12
   13
   14  7
   ```

6. 220, 234, 258, 222, 229, 226, 182, 228, 209, 180, 183, 199, 230, 210, 220, 168, 194, 164, 211, 211, 202, 222, 172, 203, 215
   Ans:
   ```
   16  4 8
   17  2
   18  0 2 3
   19  4 9
   20  2 3 9
   21  0 1 1 5
   22  0 0 2 2 6 8 9
   23  0 4
   24
   25  8
   ```

7. 7428, 7394, 7198, 7122, 7575, 7299, 7169, 7195
   Ans:
   ```
   71  22 69 95 98
   72  99
   73  94
   74  28
   75  75
   ```

# FAQ 6. How Do I Make a Histogram?

A **histogram** is a graphical representation of the distribution of data. The values are represented by vertical bars.

The width of each bar in a histogram is the same as the width of every other bar. There should not be any space between the bars. The width of a single bar is determined by dividing the **width of the** $x$**-axis** into the total **desired number of bars.**

There is no correct **number of bars.** As a general rule of thumb, you should have **at least five bars** and **no more than 15 bars.**

Each bar represents the total number (or count) of items in the data that falls between a specific range of values (along the $x$-direction). The range is sometimes called a **bin** or **bucket**, because it is thought to collect all the data points that fall between the left and right end points of the interval along the $x$-axis that the vertical bar represents. The heights of the bars are sometimes called the **bin counts.**

Here is a general procedure for drawing a histogram.

1. Determine the overall range of the data. The range is the interval [**min, max**] where **min** is the smallest number in the data set, and **max** is the largest number in the data set.
   **Example 6.1.** Make a histogram of the data in shown in table 6.1. In table 6.1, the largest value is 1195.5 (New Jersey) and the lowest value is 1.2 (Alaska). The overall data range is [1.2 to 1195.5]

2. Round off the total data range and divide into a round number of bins.
   a) Let $B$ be a round number that is as large as **max** or larger.
      **(Continuation of example 6.1.)** Let $B = 1200$.
   b) Let $A$ be a round number that is as small as **min** or smaller.
      **(Continuation of example 6.1.)** Let $A = 0$
   c) Let $n$ be the number of bars you want to have on your plot.
   d) The bin width is $W = (B - A)/n$.
      **(Continuation of example 6.1.)** If we let $n = 12$ then $W = (1200 -$

0)/12 = 100.

e) By trial an error, modify $A$, $B$, and/or $n$, as necessary, until $W$ is a nice, round number.

(**Continuation of example 6.1.**) I think 100 is nice.

A nice number in the short procedure above is something that is aesthetically pleasing. For example, $A = 50$, $B = 100$, $n = 10$, gives $W = 5$, which is nice. But $A = 40$, $B = 95$, $n = 10$, gives $W = 4.5$, which is not so nice. **There is no right answer as to what is nice.**

Table 6.1.: Population density of each of the 50 US states, in persons per square mile. The data was taken from the United States Census. (See example 6.1.)

| State | Density | State | Density | State | Density |
|---|---|---|---|---|---|
| Alabama | 94.4 | Louisiana | 104.9 | Ohio | 282.3 |
| Alaska | 1.2 | Maine | 43.1 | Oklahoma | 54.7 |
| Arizona | 56.3 | Maryland | 594.8 | Oregon | 39.9 |
| Arkansas | 56 | Massachusetts | 839.4 | Pennsylvania | 283.9 |
| California | 239.1 | Michigan | 174.8 | Rhode Island | 1018.1 |
| Colorado | 48.5 | Minnesota | 66.6 | South Carolina | 153.9 |
| Connecticut | 738.1 | Mississippi | 63.2 | South Dakota | 10.7 |
| Delaware | 460.8 | Missouri | 87.1 | Tennessee | 153.9 |
| Florida | 350.6 | Montana | 6.8 | Texas | 96.3 |
| Georgia | 168.4 | Nebraska | 23.8 | Utah | 33.6 |
| Hawaii | 211.8 | Nevada | 24.6 | Vermont | 67.9 |
| Idaho | 19 | New Hampshire | 147 | Virginia | 202.6 |
| Illinois | 231.1 | New Jersey | 1195.5 | Washington | 101.2 |
| Indiana | 181 | New Mexico | 17 | West Virginia | 77.1 |
| Iowa | 54.5 | New York | 411.2 | Wisconsin | 105 |
| Kansas | 34.9 | North Carolina | 196.1 | Wyoming | 5.8 |
| Kentucky | 109.9 | North Dakota | 9.7 | | |

3. Make a list of the **left end points** of each of the **bins** (or **buckets**) by dividing the total $x$-axis into equal sub-intervals. Using the final values of $A$, $B$, $n$ and $W$ determined above, these are

$$A$$
$$A + W$$
$$A + 2W$$
$$\cdots$$
$$A + (n - 1)W$$

(**Continuation of example 6.1.**) Since $A = 0$, $W = 100$, and $n = 12$, the left end points are 0, 100, 200, 300, ..., 1100. Note that 1200 is not a left endpoint as it is the right endpoint of the $x$-axis.

4. Write the list of the **right end points** of each bin. The right end point of each bin is the left end point of the next bin. The right end point of the largest bin the right end point of the $x$-axis.
   (**Continuation of example 6.1.**)   The corresponding right end points are 100, 200, 300, ..., 1200. The bins are summarized in the first three columns of table 6.2.

Table 6.2.: Bin assignments for population data example.

| Bin Number | Left End Point | Right End Point | Accept Point If: |
|:---:|:---:|:---:|:---:|
| 1 | 0 | 100 | $0 \leqslant x < 100$ |
| 2 | 100 | 200 | $100 \leqslant x < 200$ |
| 3 | 200 | 300 | $200 \leqslant x < 300$ |
| 4 | 300 | 400 | $300 \leqslant x < 400$ |
| 5 | 400 | 500 | $400 \leqslant x < 500$ |
| 6 | 500 | 600 | $500 \leqslant x < 600$ |
| 7 | 600 | 700 | $600 \leqslant x < 700$ |
| 8 | 700 | 800 | $700 \leqslant x < 800$ |
| 9 | 800 | 900 | $800 \leqslant x < 900$ |
| 10 | 900 | 1000 | $900 \leqslant x < 1000$ |
| 11 | 1000 | 1100 | $1000 \leqslant x < 1100$ |
| 12 | 1100 | 1200 | $1100 \leqslant x < 1200$ |

5. Examine each entry $x$ in the data set. Compare its value with the left and right end points of each bin. There will be precisely one bin for which

$$(\text{left endpoint}) \leqslant x < (\text{right end point})$$

Record (i.e., write down) the bin number.
(**Continuation of example 6.1.**) We have tabulated the assignments of each state's population density to the appropriate bin in table 6.3.

6. Count the number of items that were marked for each bin, tabulate them.
   (**Continuation of example 6.1.**) Examining the bin assignments in table 6.3, we see that 25 states have been assigned to bin 1, 11 states have been assigned to bin 2, 6 states have been assigned to 3, and so forth. The results are summarized in table 6.4.

7. Draw the histogram. Draw $n$ rectangles. For the $i^{th}$ rectangle,

   - The left edge is at $x = $ left endpoint of bin $i$
   - The right edge is at $x = $ right endpoint of bin $i$
   - The bottom is at $y = 0$.
   - The top of is at $y = $ number of counts for bin $i$

Table 6.3.: Each state has been assigned to a bin, defined by table 6.2, according to its population density as given in table 6.1.

| State | Density | Bin | State | Density | Bin |
|-------|---------|-----|-------|---------|-----|
| Alabama | 94.4 | 1 | Montana | 6.8 | 1 |
| Alaska | 1.2 | 1 | Nebraska | 23.8 | 1 |
| Arizona | 56.3 | 1 | Nevada | 24.6 | 1 |
| Arkansas | 56 | 1 | New Hampshire | 147 | 2 |
| California | 239.1 | 3 | New Jersey | 1195.5 | 12 |
| Colorado | 48.5 | 1 | New Mexico | 17 | 1 |
| Connecticut | 738.1 | 8 | New York | 411.2 | 5 |
| Delaware | 460.8 | 5 | North Carolina | 196.1 | 2 |
| Florida | 350.6 | 4 | North Dakota | 9.7 | 1 |
| Georgia | 168.4 | 2 | Ohio | 282.3 | 3 |
| Hawaii | 211.8 | 3 | Oklahoma | 54.7 | 1 |
| Idaho | 19 | 1 | Oregon | 39.9 | 1 |
| Illinois | 231.1 | 3 | Pennsylvania | 283.9 | 3 |
| Indiana | 181 | 2 | Rhode Island | 1018.1 | 11 |
| Iowa | 54.5 | 1 | South Carolina | 153.9 | 2 |
| Kansas | 34.9 | 1 | South Dakota | 10.7 | 1 |
| Kentucky | 109.9 | 2 | Tennessee | 153.9 | 2 |
| Louisiana | 104.9 | 2 | Texas | 96.3 | 1 |
| Maine | 43.1 | 1 | Utah | 33.6 | 1 |
| Maryland | 594.8 | 6 | Vermont | 67.9 | 1 |
| Massachusetts | 839.4 | 9 | Virginia | 202.6 | 3 |
| Michigan | 174.8 | 2 | Washington | 101.2 | 2 |
| Minnesota | 66.6 | 1 | West Virginia | 77.1 | 1 |
| Mississippi | 63.2 | 1 | Wisconsin | 105 | 2 |
| Missouri | 87.1 | 1 | Wyoming | 5.8 | 1 |

(**Continuation of example 6.1.**) The histogram is illustrated in fig. 6.1

Figure 6.1 Histogram of the population density data, table 6.1.

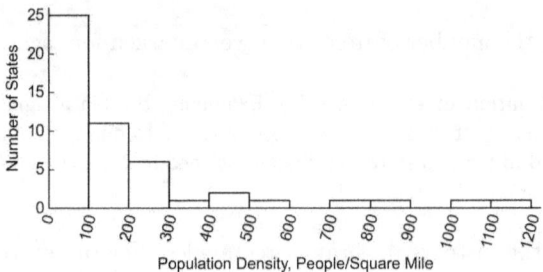

**Technology Example 6.2 (TI-84).** Plot a histogram of the data in table 6.1 using a TI-84.

Table 6.4.: Final histogram results for the population density data.

| Bin Number | Left | Right | Count |
|:---:|:---:|:---:|:---:|
| 1 | 0 | 100 | 25 |
| 2 | 100 | 200 | 11 |
| 3 | 200 | 300 | 6 |
| 4 | 300 | 400 | 1 |
| 5 | 400 | 500 | 2 |
| 6 | 500 | 600 | 1 |
| 7 | 600 | 700 | 0 |
| 8 | 700 | 800 | 1 |
| 9 | 800 | 900 | 1 |
| 10 | 900 | 1000 | 0 |
| 11 | 1000 | 1100 | 1 |
| 12 | 1100 | 1200 | 1 |

1. Hit [stat] ⟩⟩ EDIT ⟩⟩ [Edit] followed by [enter].

2. Navigate to the L1 column and enter the data, following each number by [enter]. Hit [94.4] [enter] [1.2] [enter] ⋯ [5.8] [enter].

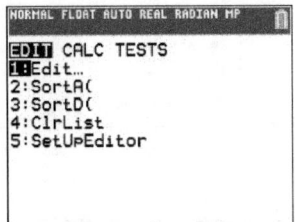 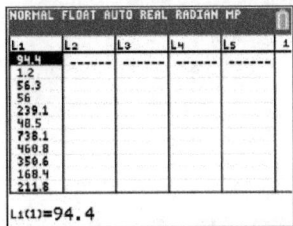

3. Press [zoom] ⟩⟩ ZOOM ⟩⟩ ZoomStat (or [zoom] [9])

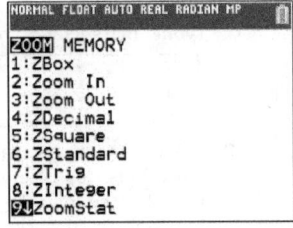 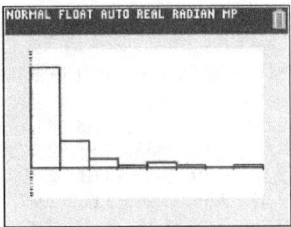

**Technology Example 6.3 (R).** Plot a histogram of the data in table 6.1 using R.

We can do this with the function **hist**. First, we have to combine the data into a vector using the **c** function.

```
> popdata=c(94.4,1.2,56.3,56,239.1,48.5,738.1,
+ 460.8,350.6,168.4,211.8,19,231.1,181,54.5,
+ 34.9,109.9,104.9,33.1,594.8,839.4,174.8,66.6,
+ 63.2,87.1,6.8,23.8,24.6,147,1195.5,17,411.2,
+ 196.1,9.7,282.3,54.7,39.9,283.9,1018.1,153.9,
+ 96.3,33.6,67.9,202.6,101.2,77.1,105,5.8)
> hist(popdata)
```

The **+** sign at the beginning of each line is a prompt that is automatically printed by RStudio whenever you hit [enter] without completing the input; it indicates that you have not completed your input for the current line. It knows you are not finished because the right parenthesis is missing.

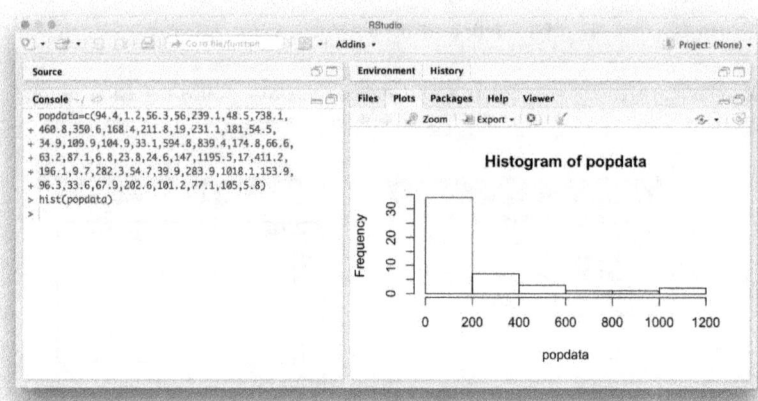

# Exercises

1. Make histograms of the following data:

2.80, 2.20, 2.33, 3.08, 3.02, 3.24, 3.22, 3.10, 2.88, 1.88, 2.73, 2.88, 2.19, 3.29, 3.09, 3.01, 2.80, 2.80, 2.47, 2.65, 3.22, 2.63, 2.92, 2.54, 2.35, 2.62, 3.15, 3.12, 2.03, 2.73, 2.63, 3.06, 3.21, 2.48, 3.18, 2.93, 2.87, 3.26, 2.90, 3.15, 3.08, 3.00, 3.17, 3.16, 3.08, 2.96, 2.05, 2.39, 3.05, 3.02

Use (a) 5 ins; (b) 8 bins; (c)

10 bins; and (d) 20 bins. Label your axes clearly. answer: (your plots could look different)

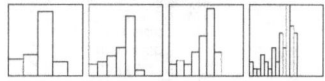

2. Make histograms using (a) 5 ins; (b) 8 bins; (c) 10 bins; and (d) 20 bins, for the following data set Label your axes clearly.

3.02, 2.30, 2.37, 2.77, 3.55, 3.04, 3.30, 1.99, 2.64, 1.89, 3.08, 3.07, 2.84, 3.36, 2.09, 2.89, 3.17, 2.54, 2.64, 3.26, 2.68, 3.30, 3.40, 2.58,

2.55, 2.82, 2.70, 2.05, 1.80, 3.31,
2.39, 1.95, 2.04, 2.74, 2.34, 2.96,
1.93, 2.91, 3.77, 2.20, 3.42, 2.23,
2.59, 2.23, 2.69, 2.31, 2.34, 2.60,
3.44, 3.52

Answer (your plots could look different)

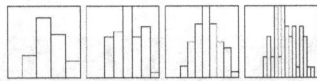

3. Plot a histogram of the following data.

25, 18, 24, 10, 35, 40, 31, 21, 24,
32, 24, 27, 37, 33, 20, 20, 28, 32,
15, 24, 22, 19, 29, 25, 21, 31, 35,
32, 23, 25, 17, 29, 30, 19, 15, 14,

34, 26, 30, 30, 32, 17, 20, 32, 25,
34, 20, 33, 28, 25, 13, 39, 21, 22,
25, 33, 36, 34, 15, 8, 11, 14, 25, 39,
20, 35, 23, 21, 53, 17, 30, 23, 19,
29, 54, 19, 27, 38, 18, 43, 35, 20,
17, 27, 24, 43, 24, 9, 32, 31, 30, 31,
19, 31, 35, 15, 23, 29, 16, 22

a) What is the data range?
b) How many bins should you use?
c) How large will each bin be?
d) Determine how many data point go into each bin.
e) Carefully draw the histogram.

# FAQ 7. How Do I Find the Median?

When a **list of numbers** is sorted numerically from smallest to largest, the **median** is the number in the middle. If there are an even number of items in the list, so that no number is precisely in the middle, then the median is the average of the two numbers closest to the middle of the list. **In a continuous distribution**, the median is the equal-area point.

## Finding the Length of the List

To find the **length** of a list of numbers, count the numbers. The length of the list is the total **count** of the items in the list. Call this number $n$.

**Example 7.1.** The list:

$$7, \quad 6, \quad 13, \quad 4$$

has a length of $n = 4$, while the list

$$7, \quad 6, \quad 13, \quad 4, \quad 9, \quad 12, \quad 16, \quad 3$$

has a length of $n = 8$.

**Example 7.2.** In the following we show two rows of numbers. The first row is a counter starting at 1 and increasing by 1 from left to right. The second row contains the same sequence of numbers as in the second list from example 7.1

above.

$$\begin{array}{cccccccc} 1 & 2 & 3 & 4 & 5 & 6 & 7 & 8 \\ \downarrow & \downarrow & \downarrow & \downarrow & \downarrow & \downarrow & \downarrow & \downarrow \\ 7, & 6, & 13, & 4, & 9, & 12, & 16, & 3 \end{array}$$

The arrows help us to visualize which number is first, which number is second, which number is third, and so forth.

## Finding the Median of a List of Numbers

1. Sort the list in numerical order, from smallest to largest.
2. If the length of the list is odd, the median is number in the middle of the list.
3. If the length of the list is even, the median is average of the two numbers in the middle.

If the length of a list is odd, like 3, 5, 7, 9, ..., we say that the list has an **odd length.**

If the length of a list is even, like 4, 6, 8, 10, 12,..., we say that the list has an **even length.**

**Technology Example 7.3 (R).** Find the length of the list using R:
$$22, 33, 31, 36, 23, 25, 23, 20, 35, 33, 33$$
We can use the R function **length** for this.

```
> x=c(22,33,31,36,23,25,23,20,35,33,33)
> length(x)
[1] 11
```

Thus the length of the list is 11.

The procedures for finding the medians of even-length lists ad odd-length lists are slightly different.

## Median of a list with Odd Length

The median of a list of numbers with odd length is the number in the middle of the list, *after it has been sorted in order numerically,* from smallest to largest. This number is at position $(n + 1)/2$ of the *sorted* list, where $n$ is the length of the list.

## Median of an Odd-Length List

1. Find the length $n$ of the list.

   **Example 7.4.** Find the median of

$$1, \quad 6, \quad 2, \quad 1, \quad 12, \quad 25, \quad 23, \quad 14, \quad 13, \quad 8, \quad 20$$

   Counting the elements of the list from left to right,

| 1 | 6 | 2 | 1 | 12 | 25 | 23 | 14 | 13 | 8 | 20 | ← data |
|---|---|---|---|----|----|----|----|----|---|----|--------|
| ↑ | ↑ | ↑ | ↑ | ↑  | ↑  | ↑  | ↑  | ↑  | ↑ | ↑  | |
| 1 | 2 | 3 | 4 | 5  | 6  | 7  | 8  | 9  | 10 | 11 | ← counter |

   Hence the length of the list is $n = 11$.

2. Verify that $n$ is odd. If it is even, skip to the section below titled *Median of an Even-Length List*.

   (**Continuation of example 7.4.**) Since $n = 11$, which is odd, we have an odd list, so we may continue with step 3.

3. Sort the list in increasing order. If you are doing this by hand, it is probably easier to do a stem plot first (see chapter 5). If you do a stem plot, you can just read the sorted numbers directly off the plot. Remember to keep all repeats that are present in the data.

   (**Continuation of example 7.4.**) A stem plot of this data set looks like this:

$$
\begin{array}{c|l}
0 & 1\ 1\ 2\ 6\ 8 \\
1 & 2\ 3\ 4 \\
2 & 0\ 3\ 5
\end{array}
$$

   Reading off the data in order gives the sorted list:

$$1, \quad 1, \quad 2, \quad 6, \quad 8, \quad 12, \quad 13, \quad 14, \quad 20, \quad 23, \quad 25$$

4. The center of an odd list is at location $(n + 1)/2$, so count to that location.

   (**Continuation of example 7.4.**) In the example, $n = 11$, so $n + 1 = 12$. The center position is $(n + 1)/2 = 12/2 = 6$

| 1 | 1 | 2 | 6 | 8 | 12 | 13 | 14 | 20 | 23 | 25 | ← sorted data |
|---|---|---|---|---|----|----|----|----|----|----|----|
| ↑ | ↑ | ↑ | ↑ | ↑ | ↑  | ↑  | ↑  | ↑  | ↑  | ↑  | |
| 1 | 2 | 3 | 4 | 5 | 6  | 7  | 8  | 9  | 10 | 11 | ← position |

In the example, the number in the sixth position is 12 (outlined). This number is the median, $M = 12$.

You could also find the median by counting numbers over and down in the stem plot. First count across, then count down.

(**Continuation of example 7.4.**) The median is circled below:

$$
\begin{array}{c|l}
0 & 1\ 1\ 2\ 6\ 8 \\
1 & ②\ 3\ 4 \\
2 & 0\ 3\ 5
\end{array}
$$

Start with the first row and count across the right until you get to the sixth item $((n + 1)/2 = 6)$. This is the 2 on the second row. Since the second row has a stem of 10, the median is 12.

# Median of an Even-Length List

## Median of a list with Even Length

The median of a list of numbers with even length is the average of the two number in the middle of the list, *after the list has been sorted in order numerically*, from smallest to largest. These two numbers are at positions $n/2$ and $(n/2)+1$ of the *sorted* list, where $n$ is the length of the list.

1. Find the length $n$ of the list.

   **Example 7.5.** Find the median of the following list:

   $$2, \quad 18, \quad 5, \quad 1, \quad 19, \quad 17, \quad 3, \quad 15, \quad 16, \quad 22, \quad 4, \quad 8$$

   We begin by counting the number of elements in the list.

   | 2 | 18 | 5 | 1 | 19 | 17 | 3 | 15 | 16 | 22 | 4 | 8 | ← data |
   |---|----|---|---|----|----|---|----|----|----|---|---|--------|
   | ↑ | ↑ | ↑ | ↑ | ↑ | ↑ | ↑ | ↑ | ↑ | ↑ | ↑ | ↑ | |
   | 1 | 2 | 3 | 4 | 5 | 6 | 7 | 8 | 9 | 10 | 11 | 12 | ← counter |

   We see that the length of the list is $n = 12$.

2. Verify that $n$ is even. If it is odd, go to the section titled *Median of an Odd-Length List.*

(**Continuation of example 7.5.**) The length of the list is $n = 12$, which is an even number. Therefore we have an even list, and we may continue with step 3.

3. Sort the list in increasing order. It you are sorting the list by hand then it is generally easier to do a stem plot first (see chapter 5) first. Remember to keep any repeats.

(**Continuation of example 7.5.**) A stem plot of this data set looks like this:

$$
\begin{array}{c|l}
0 & 123458 \\
1 & 56789 \\
2 & 2
\end{array}
$$

There are no repeats in this example; the sorted list looks like this:

$$1, \quad 2, \quad 3, \quad 4, \quad 5, \quad 8, \quad 15, \quad 16, \quad 17, \quad 18, \quad 19, \quad 22$$

4. The center of the list falls between position $n/2$ and $(n/2) + 1$.

(**Continuation of example 7.5.**) Since $n = 12$, the middle two numbers are at position, $n/2 = 12/2 = 6$ and location $(n/2) + 1 = 6 + 1 = 7$. The median falls *between* position 6 and 7:

| 1, | 2, | 3, | 4, | 5, | 8, | 15 | 16, | 17, | 18, | 19, | 22 | ← data |
|----|----|----|----|----|----|----|-----|-----|-----|-----|----|--------|
| ↑ | ↑ | ↑ | ↑ | ↑ | ↑ | ↑ | ↑ | ↑ | ↑ | ↑ | ↑ | |
| 1 | 2 | 3 | 4 | 5 | 6 | 7 | 8 | 9 | 10 | 11 | 12 | ← position |

In the example, the numbers in the sixth and seventh position are 8 and 15. We can also find the two center values by carefully counting locations into the stem plot.

(**Continuation of example 7.5.**) Here we locate the median by counting locations in the stem plot.

$$
\begin{array}{c|l}
0 & 12345\,(8) \\
1 & (5)\,6789 \\
2 & 2
\end{array}
$$

5. The median is the average of the two numbers in the center.

---

(**Continuation of example 7.5.**)  Therefore the median is

$$M = (8 + 15)/2 = 23/2 = 11.5$$

**Double-Checking your Answer.** Sort the list and count the number of items that are on either side of the median. The number of items that are greater than the median should be equal to the number of items that are less than the median. If it does not split the data in this way, then the number is not the median.

(**Continuation of example 7.5.**) To verify that 11.5 split the data correctly we observe that six numbers are less than or equal to 11.5 (in this case they are all less than the median: 1, 2, 3, 4, 5, 8) and six numbers are greater than or equal 11.5 (in this case, they are all strictly greater than the median: 15, 16, 17, 18, 19 22). Thus it splits the data in two equal-sized sets[1]

**Technology Example 7.6 (TI-84).** Find the median using a TI-84 calculator:
$$22, 33, 31, 36, 23, 25, 23, 20, 35, 33, 33$$

1. Enter the numbers into a list such as L1, such as in example 6.2. Select stat EDIT ≫ Edit enter .

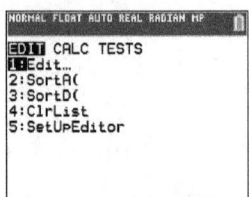 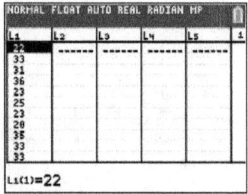

2. Navigate to the L1 column using the arrow keys.
3. Enter each number followed by enter . In this case, you would type in: 22 enter 33 enter ⋯ 33 enter 33 enter
4. Enter 2nd list followed by MATH ≫ median (i.e., scroll down to median in the MATH menu). Then press enter .
5. When the **median** prompt appears, select the list **L1** by hitting 2nd L1 ) enter . Read off the median (31, in this case).

---

[1]Note: this does not actually tell us that we have found the correct median. It will only tell us if we have split the data set at the wrong location. So it is not a foolproof test. It is presented here because it is an easy check.

---

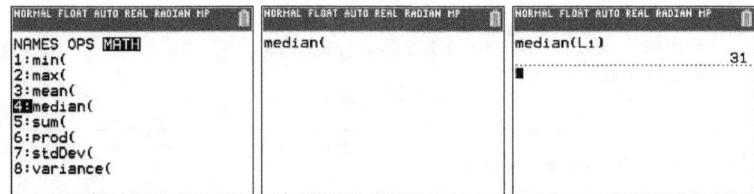

6. Alternatively, the median can is also obtained as one item of descriptive statistics in one-variable statistics mode. Instead of entering list/math mode in step 4, type [stat] [CALC ⟩ 1-Var Stats] [enter] to get to the 1-variable statistics menu.
7. On the first line select [List ⟩ L1]. This selects the data in list L1.
8. Navigate to [Calculate:] and press [enter].

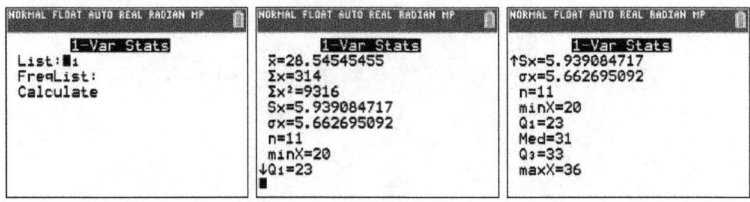

The arrow at the bottom left means some data is not displayed. Scroll to the second page using the arrow keys. The median is labeled **Med**. The other fields in this output are described in table 7.1.

Table 7.1.: Description of fields in the results of the one-variable calculate mode for the TI-84.

| Label | Definition |
|---|---|
| $\bar{x}$ | mean (ch. 13) |
| $\Sigma x$ | sum of all numbers |
| $\Sigma x^2$ | sum of squares |
| Sx | standard deviation (ch. 15) |
| $\sigma x$ | population standard deviation |
| n | count of data items |
| minX | smallest data item (ch. 9) |
| Q1 | $Q_1$ (ch. 8) |
| Med | Median (ch. 7) |
| Q3 | $Q_3$ (ch. 8) |
| maxX | Largest data item (ch. 9) |

**Technology Example 7.7 (R).** Find the median using R:
$$22, 33, 31, 36, 23, 25, 23, 20, 35, 33, 33$$
Use the function **median**.

```
> x=c(22,33,31,36,23,25,23,20,35,33,33)
> median(x)
[1] 31
```

Thus the median is 31.

**Technology Example 7.8 (Spreadsheet).** Find the median of the following data using a spreadsheet:

$$22, 33, 31, 36, 23, 25, 23, 20, 35, 33, 33$$

1. Enter the numbers in a column in the spreadsheet.
2. In any cell type the expression =median(.

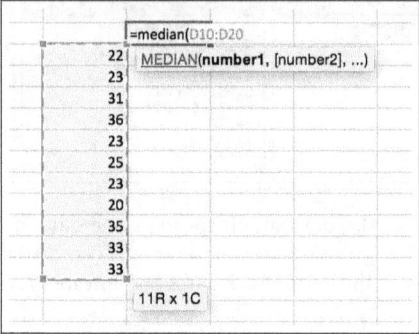

3. Place your cursor just to the right of the right parenthesis, then use the mouse to select the data items you want in the column of data. The addresses of the data items will appear.
4. Type a ) and then enter. The answer will appear in the cell.

# Exercises

Find the medians of the following lists.

1. 78, 48, 52, 71, 47, 56, 52, 54, 39, 49, 52, 34, 43, 75, 46 (ans: 52)
2. 35, 45, 53, 57, 66, 61, 49, 44, 44, 45, 47, 55 (ans: 48)
3. 29, 52, 48, 66, 51, 44, 62 (ans: 51)
4. 83, 56, 42, 56, 99, 118, 147, 102, 112, 108, 52, 110, 86, 50, 54 (ans: 86)
5. 6, 8, 10, 7, 13, 13, 11, 8, 10, 9 (ans: 9.5)
6. 9, 12, 10, 10, 9, 15, 7, 11, 8, 13, 12, 10, 12 (ans: 10)
7. 10, 11, 11, 13, 11, 8, 12, 11, 7, 11, 13, 7 (ans: 11)
8. 42.2, 49.6, 43.3, 64.8, 53.7, 42.3, 55.5 (ans: 49.6)
9. 73.7, 71.4, 70.8, 72.2, 73.4, 74.0, 72.3 (ans: 72.3)
10. 55.9, 54.4, 52.9, 47.6, 44.2, 52.2, 54.3, 53.2 (ans: 53.05)

# FAQ 8. How Do I Find the Quartiles?

The **quartiles** are numbers that divide a data set into four **quarters**. The numbers $Q_1$, $Q_2$, $Q_3$, and $Q_4$ refer to the upper end (larger value) of each of the four quarters. $Q_2$ is the same as the median. $Q_4$ is the same as the maximum (largest) number in the data set. Sometimes the quartiles are referenced without subscripts, e.g, as Q1 or Q3.

---

### Quartile

A **quartile** is one-fourth of the data set, counted by number of data points. Quartiles are arranged by values, from the lowest quartile, which contains the one-quarter of the data points that have the smallest values, to the highest quartile, which contain the one-quarter of the data points with the largest values.

---

The numbers $Q_1$, $M$ and $Q_3$ are the boundaries between quartiles. The number $Q_1$ is boundary between the lower two quartiles, and the number $Q_3$ is the boundary between the upper two quartiles.

Figure 8.1.: $Q_1$, $M$, and $Q_3$ separate data into fourths.

| $Q_1$ | $M$ | $Q_3$ | |
|---|---|---|---|
| first quartile | second quartile | third quartile | fourth quartile |
| 25% | 25% | 25% | 25% |

One quarter (1/4, or 25%) of the data values are smaller than $Q_1$, and one quarter (1/4 or 25% of the data values are larger than $Q_3$. The median (chapter 7) lies at the boundary between the middle two quartiles. Half of the data (1/2 or 50%) lies in the two center quartiles, between $Q_1$ and $Q_3$.

## Finding the Quartiles: Odd-Sized Data Sets

1. Count the number of items in the data set to verify that there are an odd number of items in the data set. (Otherwise, go to the section *Finding the Quartiles: Even-Sized Data Sets*.)

**Example 8.1.** Find the quartiles:.

$$1\ \ 2\ \ 3\ \ 4\ \ 5\ \ 6\ \ 7\ \ 8\ \ 9\ \ 10\ 11\ 12\ 13\ 14\ 15 \leftarrow \text{counter}$$
$$\downarrow\ \downarrow\ \downarrow\ \downarrow\ \downarrow\ \downarrow\ \downarrow\ \downarrow\ \downarrow\ \downarrow\ \downarrow\ \downarrow\ \downarrow\ \downarrow\ \downarrow$$
$$57, 46, 35, 44, 53, 49, 53, 41, 50, 45, 41, 38, 39, 51, 46 \leftarrow \text{data values}$$

This data set has 15 elements. Since 15 is odd, we continue with step 2.

**Example 8.2.** Find the quartiles :

```
 1   2   3   4   5   6   7   8   9  10  11  12  13  14  15  16  17 ← counter
 ↓   ↓   ↓   ↓   ↓   ↓   ↓   ↓   ↓   ↓   ↓   ↓   ↓   ↓   ↓   ↓   ↓
52, 48, 66, 51, 44, 62, 53, 42, 37, 43, 60, 67, 79, 61, 65, 63, 41 ← data values
```

This data set has 17 elements. Since 17 is odd, we continue with step 2.

2. Sort the data and find the median (ch. 7).

   **(Continuation of example 8.1.)** There are 15 data items. The center is in location $(15 + 1)/2 = 8$, as illustrated here.

```
 1   2   3   4   5   6   7    8    9   10  11  12  13  14  15 ← counter
 ↓   ↓   ↓   ↓   ↓   ↓   ↓    ↓    ↓   ↓   ↓   ↓   ↓   ↓   ↓
35, 38, 39, 41, 41, 44, 45, | 46, | 46, 49, 50, 51, 53, 53, 57 ← sorted data
```

   The median is $M = 46$.

   **(Continuation of example 8.2.)** There are 17 data items. The center location is $(17 + 1)/2 = 9$.

```
 1   2   3   4   5   6   7   8    9    10  11  12  13  14  15  16  17 ← counter
 ↓   ↓   ↓   ↓   ↓   ↓   ↓   ↓    ↓    ↓   ↓   ↓   ↓   ↓   ↓   ↓   ↓
37, 41, 42, 43, 44, 48, 51, 52, | 53, | 60, 61, 62, 63, 65, 66, 67, 79 ← sorted data
```

   The median is $M = 53$.

3. The median divides the original data set into two smaller data sets. Find the medians of each of these smaller data sets (ch. 7). If the original data set has an odd count, do not use the median when considering either of these data sets. If the original data set has an even count and the median was calculated as an average, we will included the two numbers used to calculated the median in their respective half-sets.

   a) If there are an odd number of items on either side of the median (excluding the median), take the number in the center of the left as $Q_1$ and the number in the middle of the right as $Q_3$.

   **(Continuation of example 8.1.)** This data set has an odd number of items – seven – on either side of the median. We find the quartiles by finding the median of each of these 7-number sets.

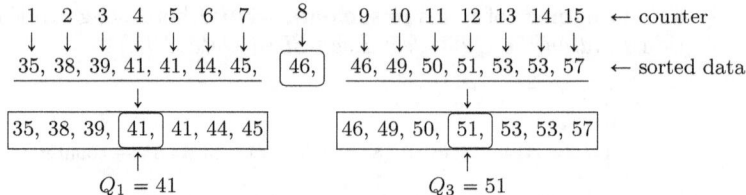

---

FAQ 8.   How Do I Find the Quartiles?

**Important note (alternative method):** Some authors will include the median in each half when counting the quartiles. This is not incorrect, it is just a different method for calculating the quartiles.

(**Continuation of example 8.1.**) If the alternative method is used, the quartiles are $Q_1 = 41$ and $Q_3 = 50.5$.

**Important note (second alternative method):** Some authors will take the average of the result found using the first method and the alternative method.

(**Continuation of example 8.1.**)I f the second alternative method is used, the quartiles are $Q_1 = 41$ and $Q_3 = 50.75$.

b) If there are an even number of items on either side of the median (excluding the median), average the two items in the middle of the left-hand set to get $Q_1$ and the two items in the middle of the right hand set to get $Q_3$.

(**Continuation of example 8.2.**)This data set has an even number of items - eight - on each side of the median. The quartiles are the medians of each half. Since the halves are even in length, we have to average two numbers to get the corresponding quartile boundary.

$$Q_1 = (43 + 44)/2 = 43.5 \qquad\qquad Q_3 = (63 + 65)/2 = 64$$

(**Continuation of example 8.2.**) If the first alternate method is used, $Q_1 = 44$ and $Q_3 = 63$.

(**Continuation of example 8.2.**) If the second alternate method is used, $Q_1 = (44 + 43.5)/2 = 43.75$ and $Q_3 = (63 + 64)/2 = 63.5$.

## Finding the Quartiles: Even-Sized Data Sets

1. Count the number of items in the data set to verify that that it is even. (Otherwise, return to the section titled: *Finding the Quartiles: Odd-Sized Data Sets*).

**Example 8.3.** Find the quartiles:

```
 1   2   3   4   5   6   7   8   9  10  11  12  13  14  ←counter
 ↓   ↓   ↓   ↓   ↓   ↓   ↓   ↓   ↓   ↓   ↓   ↓   ↓   ↓
57, 46, 35, 44, 53, 49, 53, 41, 50, 45, 41, 38, 39, 51  ← data values
```

This data set has 14 items. Since 14 is an even number, we continue with the procedure in this section.

**Example 8.4.** Find the quartiles:

```
 1   2   3   4   5   6   7   8   9  10  11  12  13  14  15  16  ←counter
 ↓   ↓   ↓   ↓   ↓   ↓   ↓   ↓   ↓   ↓   ↓   ↓   ↓   ↓   ↓   ↓
52, 48, 66, 51, 44, 62, 53, 42, 37, 43, 60, 67, 79, 61, 65, 63  ←data values
```

This data set has 16 items. Since 16 is an even number, we continue with the procedure in this section.

2. Sort the data and find the median (ch. 7).

(**Continuation of example 8.3.**) This data set has 14 elements.

```
 1   2   3   4   5   6    7    8    9  10  11  12  13  14  ←counter
 ↓   ↓   ↓   ↓   ↓   ↓    ↓    ↓    ↓   ↓   ↓   ↓   ↓   ↓
35, 38, 39, 41, 41, 44, [45,] [46,] 49, 50, 51, 53, 53, 57  ←sorted data
```

$$M = (45 + 46)/2 = 45.5$$

The center is between locations 7 and 8. Therefore the median is $M = 45.5$.

(**Continuation of example 8.4.**) This data set has 16 elements.

```
 1   2   3   4   5   6   7    8    9   10  11  12  13  14  15  16  ←counter
 ↓   ↓   ↓   ↓   ↓   ↓   ↓    ↓    ↓    ↓   ↓   ↓   ↓   ↓   ↓   ↓
37, 42, 43, 44, 48, 51, 52, [53,] [60,] 61, 62, 63, 65, 66, 67, 79  ←sorted data
```

$$M = (53 + 60)/2 = 56.5$$

The center is between locations 8 and 9. Therefore the median is $M = 56.5$.

3. The median divides the original data set into two smaller data sets. Find the medians of each of these smaller data sets (ch.7). Each half-list will include one of the numbers that was used to compute the median.

a) If there are an odd number of items in either half-list, we take the number in the center of the left half-list as $Q_1$, and the number in the middle of the right half-list as $Q_3$.

(**Continuation of example 8.3.**) Since there are originally 14 data points, each half-list has seven numbers. The quartiles are the middle numbers of these seven item sub-lists:

$$Q_1 = 41 \qquad Q_3 = 51$$

The quartiles are $Q_1 = 41$ and $Q_3 = 51$.

**Important note:** (Alternate method.) Some authors do not include the two numbers used to calculate the median in the halves uses to determine the quartiles.

(**Continuation of example 8.3.**) If the alternate method is used, $Q_1 = (39 + 41)/2 = 40$ and $Q_3 = (51 + 53)/2 = 52$.

**Important note:** (Second alternate method.) Some authors will average the values found using the first method and the alternate method to determine the quartiles.

(**Continuation of example 8.3.**) If the second alternate method is used, $Q_1 = (40 + 41)/2 = 40.5$ and $Q_3 = (52 + 51)/2 = 51.5$.

b) If there are an even number of items on either side of the median (excluding the median), we average the two items in the middle of the left-hand set to get $Q_1$ and the two values in the middle of the right hand set to get $Q_3$.

(**Continuation of example 8.4.**) This data set has 16 elements; each half has eight elements. The quartiles are computed as the medians of these eight item sequences.

$$Q_1 = (44 + 48)/2 = 46 \qquad Q_3 = (63 + 65)/2 = 64$$

The median of each half-set is computed as the average of the two middle numbers in the same manner as the median of an even length set. Thus $Q_1 = 46$ and $Q_3 = 64$.

(**Continuation of example 8.4.**) If the first alternate method is used, $Q_1 = 44$ and $Q_2 = 65$.

(**Continuation of example 8.4.**)If the second alternate method is used, $Q_1 = (44 + 46)/2 = 45$ and $Q_2 = (65 + 64)/2 = 64.5$.

**Technology Example 8.5 (TI-84).** Find the quartiles of the following data using a TI-84:
$$22, 33, 31, 36, 23, 25, 23, 20, 35, 33, 33$$

1. Select stat EDIT $\gg$ Edit enter. This takes you to the data entry page.

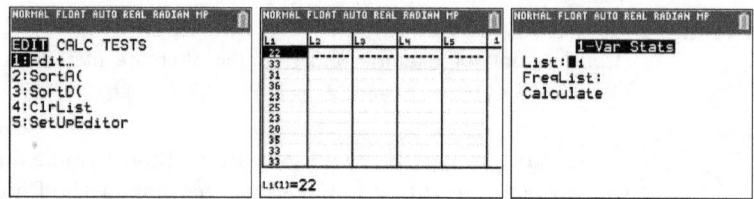

2. Navigate to the L1 column using the arrow keys.
3. Enter the numbers in the column L1, following each number by the enter. In this case, you would type in:
   22 enter 33 enter 31 enter 36 enter 23 enter 25 enter 23 enter 20 enter
   35 enter 33 enter 33 enter
   To remove (erase) an entry use the delete key.
4. Type stat CALC $\gg$ 1-Var Stats enter to get to the 1-variable statistics menu.
5. On the first line select List $\gg$ L1. This selects the data in list L1.
6. Navigate to Calculate: and press enter. The following will be displayed:

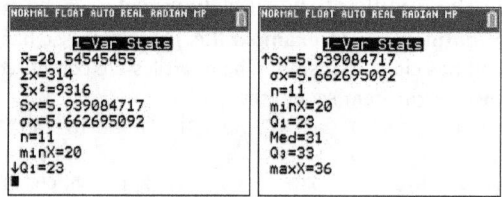

The fields in this output are described in table 7.1. The quartiles $Q_1$ and $Q_3$ are given by Q1 and Q3. We can read them off as $Q_1 = 23$ and $Q_3 = 33$.

**Technology Example 8.6 (R).** Find the quartiles of the following data using a R:
$$22, 33, 31, 36, 23, 25, 23, 20, 35, 33, 33$$
We can use the R function **quantile**.

```
> x=c(22,33,31,36,23,25,23,20,35,33,33)
> quantile(x)
```

| 0% | 25% | 50% | 75% | 100% |
|----|-----|-----|-----|------|
| 20 | 23 | 31 | 33 | 36 |

The function **quantile** returns not only $Q_1$ and $Q_3$ but also the minimum value, the median, and the maximum value. These values are sorted in numerical order as the five number summary (ch. 9). Thus $Q_1 = 23$, $M = 31$, and $Q_3 = 33$.

**Technology Example 8.7 (Spreadsheet).** Find the quartiles of the following data using a spreadsheet:

$$22, 33, 31, 36, 23, 25, 23, 20, 35, 33, 33$$

1. Enter the numbers in a column in the spreadsheet.
2. In any cell type the expression =quartile (. The syntax is **quartile (data, n)**, where **data** is a sequence of numbers in the spreadsheet and **n** is one of the numbers **0** through **4**.
3. Place your cursor just to the right of the right parenthesis, then use the mouse to select the data items you want in the column of data. The addresses of the data items will appear.
4. Type a comma.
5. Type the quartile number followed by a ) and then enter. The answer will appear in the cell.

## Exercises

Find the median M and the quartiles Q1 and Q3 of each of the following data sets.

1. 78, 48, 52, 71, 47, 56, 52, 54, 39, 49, 52, 34, 43, 75, 46 (ans: 46, 52, 56)
2. 35, 45, 53, 57, 66, 61, 49, 44, 44, 45, 47, 55 (ans: 44.5, 48, 56)
3. 29, 52, 48, 66, 51, 44, 62 (ans:44, 51, 62)
4. 83, 56, 42, 56, 99, 118, 147, 102, 112, 108, 52, 110, 86, 50, 54 (ans: 54, 86, 110)
5. 6, 8, 10, 7, 13, 13, 11, 8, 10, 9 (ans: 8, 9.5, 11)
6. 9, 12, 10, 10, 9, 15, 7, 11, 8, 13, 12,

10, 12 (ans: 9, 10, 12)
7. 10, 11, 11, 13, 11, 8, 12, 11, 7, 11,
   13, 7 (ans: 9, 11, 11.5)
8. 51.9, 52.3, 52.7, 50.5, 51.5, 52.3,
   52.7, 53.6 (ans: 51.7, 52.3, 52.7)
9. 22.0, 23.4, 25.8, 22.2, 22.9, 22.6,
   18.2, 22.8, 20.9, 18.0, 18.3, 19.9,
   23.0, 21.0, 22.0, 16.8, 19.4, 16.4,
   21.1, 21.1, 20.2, 22.2, 17.2, 20.3,
   21.5 (ans: 18.8, 21.1, 22.3)
10. 7428, 7394, 7198, 7122, 7575, 7299,
   7169, 7195 (ans: 7182, 7248, 7411)

11. 13.23, 11.74, 12.06, 13.91, 13.76,
    14.33, 14.27, 13.96, 13.42, 10.94,
    13.04, 13.43, 11.72, 14.44, 13.95,
    13.74, 13.22, 13.22, 12.41, 12.85,
    14.28, 12.81, 13.53, 12.57, 12.10,
    12.77, 14.09, 14.02, 11.30, 13.04,
    12.79, 13.87, 14.24, 12.43, 14.18,
    13.55, 13.41, 14.37, 13.47, 14.09,
    13.91, 13.72, 14.13, 14.13, 13.92,
    13.61, 11.37, 12.20, 13.84, 13.77
    (ans: 12.79, 13.54, 13.96)

# FAQ 9. What is the 5 Number Summary?

The 5-number summary describes the spread of a data set. It we label $Q_0$ as the smallest number of the data set, then the 5-number summary is

$$\{Q_0, Q_1, Q_2, Q_3, Q_4\}$$

where

$Q_0$ is the **minimum** value, or smallest number, in the data set.

$Q_1$ is the **first quartile**. Twenty five percent, or one fourth of the data points are smaller than $Q_1$, and 75% are larger.

$Q_2$ or $M$ is the **median** of the data set.

$Q_3$ is the **third quartile**. Seventy five percent, or three fourths of the data points are smaller than $Q_3$, and 25% are larger.

$Q_4$ is the **maximum** value, or the largest number, in the data set.

To find the five number summary we need to determine the quartiles. The procedure is as follows.

1. Find the **median** (ch. 7) and **quartiles** $Q_1$ and $Q_3$ (ch. 8).
   **Example 9.1.** Find the Five Number Summary of

   $$57, 46, 35, 44, 53, 49, 53, 41, 50, 45, 41, 38, 39, 51, 46$$

   The median is 46, $Q_1 = 41$, and $Q_3 = 51$ (from example 8.1).

2. Determine the **minimum** (smallest) and **maximum** (largest) values.

(**Continuation of example 9.1.**) The minimum is 35 and the maximum is 57, as illustrated here.

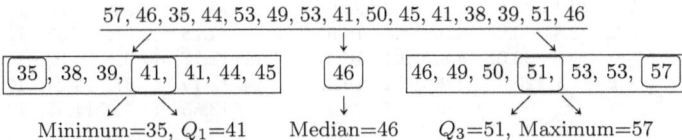

57, 46, 35, 44, 53, 49, 53, 41, 50, 45, 41, 38, 39, 51, 46

$\boxed{35}$, 38, 39, $\boxed{41}$, 41, 44, 45     $\boxed{46}$     46, 49, 50, $\boxed{51}$, 53, 53, $\boxed{57}$

Minimum=35, $Q_1$=41     Median=46     $Q_3$=51, Maximum=57

3. The **5-number summary** is then:

$$(\text{Minimum}, Q_1, \text{Median}, Q_3, \text{Maximum})$$

It does not matter how the median or quartiles $Q_1$ and $Q_3$ were determined, e.g., as the center number or as the average of two numbers. Whatever values of the median M (chapter 7) and quartiles $Q_1$ and $Q_3$ you find (chapter 8), these are the numbers you should use.

We normally enclose the five numbers is some kind of parentheses or brackets and write the five numbers in increasing numerical order. (**Continuation of example 9.1.**) In this example, the 5-number summary is $(35, 41, 46, 51, 57)$

**Technology Example 9.2 (TI-84).** Find the five number summary using a TI-84:
$$22, 33, 31, 36, 23, 25, 23, 20, 35, 33, 33$$
This is done with one-variable statistics. We have already done this for this data set in example 8.5. The five number summary can be read directly off the output screen as minxX, Q1, Med, Q3, and maxX. In this case we read off these numbers as $\{20, 23, 31, 33, 36\}$.

**Technology Example 9.3 (R).** Find the five number summary using R:
$$22, 33, 31, 36, 23, 25, 23, 20, 35, 33, 33$$
The five number summary can be found using the function **quantile**. This was already done for this data set in example 8.6.

# Exercises

Find the five number summary of each of the following data sets.

1. 78, 48, 52, 71, 47, 56, 52, 54, 39, 49, 52, 34, 43, 75, 46 Ans: [34, 46.0, 52.0, 56.0, 78]
2. 35, 45, 53, 57, 66, 61, 49, 44, 44, 45, 47, 55 Ans: [35, 44.5, 48.0, 56.0, 66]

3. 29, 52, 48, 66, 51, 44, 62 Ans: [29, 44.0, 51.0, 62.0, 66]
4. 83, 56, 42, 56, 99, 118, 147, 102, 112, 108, 52, 110, 86, 50, 54 Ans: [42, 54.0, 86.0, 110.0, 147]
5. 6, 8, 10, 7, 13, 13, 11, 8, 10, 9 Ans: [6, 8.0, 9.5, 11.0, 13]
6. 9, 12, 10, 10, 9, 15, 7, 11, 8, 13, 12, 10, 12 Ans: [7, 9.0, 10.0, 12.0, 15]
7. 10, 11, 11, 13, 11, 8, 12, 11, 7, 11, 13, 7 Ans: [7, 9.0, 11.0, 11.5, 13]
8. 51.9, 52.3, 52.7, 50.5, 51.5, 52.3,

52.7, 53.6 Ans: [50.5, 51.7, 52.3, 52.7, 53.6]

9. 22.0, 23.4, 25.8, 22.2, 22.9, 22.6, 18.2, 22.8, 20.9, 18.0, 18.3, 19.9, 23.0, 21.0, 22.0, 16.8, 19.4, 16.4, 21.1, 21.1, 20.2, 22.2, 17.2, 20.3, 21.5 Ans: [16.4, 18.85, 21.1, 22.4, 25.8]

10. 7428, 7394, 7198, 7122, 7575, 7299, 7169, 7195 Ans: [7122, 7182, 7249, 7411, 7575]

11. 13.23, 11.74, 12.06, 13.91, 13.76,

14.33, 14.27, 13.96, 13.42, 10.94, 13.04, 13.43, 11.72, 14.44, 13.95, 13.74, 13.22, 13.22, 12.41, 12.85, 14.28, 12.81, 13.53, 12.57, 12.10, 12.77, 14.09, 14.02, 11.30, 13.04, 12.79, 13.87, 14.24, 12.43, 14.18, 13.55, 13.41, 14.37, 13.47, 14.09, 13.91, 13.72, 14.13, 14.13, 13.92, 13.61, 11.37, 12.20, 13.84, 13.77 Ans: [10.94, 12.79, 13.54, 13.96, 14.44]

# FAQ 10. What is the IQR?

The **Inter-quartile Range** (IQR) is the spread of the central fifty percent of the sorted data. It is sometimes used, via the so-called **1.5 × IQR rule**, to identify potential outliers.

To find the IQR of a distribution:

1. Find the quartiles $Q_1$ an $Q_3$ of the distribution (chapter 8).

2. Calculate $IQR = Q_3 - Q_1$

**Example 10.1.** Find the IQR:

$$88, 46, 9, 44, 53, 49, 76, 41, 50, 45, 41, 38, 39, 51, 46$$

Following the procedures give in chapters 7 and 8, we find that the median is 46, $Q_1 = 41$ and $Q_3 = 51$. This is illustrated here:

$88, 46, \ 9, \ 44, 53, 49, 76, 41, 50, 45, 41, 38, 39, 51, 46$ (unsorted data)

$9, \ 38, 39, 41, 41, 44, 45, 46, 46, 49, 50, 51, 53, 76, 88$ (sorted data)

| 9 38 39 | 41 | 41 44 45 | | 46 | | 46 49 50 | 51 | 53 76 88 |

$Q_1 = 41$  Median=46  $Q_3 = 51$

IQR = 51-41 =10

Therefore the IQR is 10.

> **Inter-quartile Range (IQR)**
>
> The **Inter-quartile Range** or **IQR** of a distribution is
> $$IQR = Q_3 - Q_1$$

**Technology Example 10.2 (TI-84).** Find the IQR using a TI-84:

$$22, 33, 31, 36, 23, 25, 23, 20, 35, 33, 33$$

1. Find the 5-number summary using the TI-84. This was done in example 9.2. The result was $\{20, 23, 31, 33, 36\}$.
2. Since $Q_1 = 23$ and $Q_3 = 33$, The IQR is $Q_3 - Q_1 = 33 - 23 = 10$.

**Technology Example 10.3 (R).** Find the IQR using R:

$$22, 33, 31, 36, 23, 25, 23, 20, 35, 33, 33$$

Use the function **quantile** to find the quartiles $Q_1$ and $Q_3$ (see example 8.6).

```
> x=c(22,33,31,36,23,25,23,20,35,33,33)
> quantile(x)
   0%   25%   50%   75%  100%
   20    23    31    33    36
```

Since $Q_3 = 33$ and $Q_1 = 23$, the IQR is 10.

# FAQ 11. What is the IQR Rule for Outliers?

The **1.5 × IQR Rule** is sometimes used to identify potential outliers. Outliers are data points that are far away from the rest of the data, or points that do not match the rest of the distribution in terms its overall shape or variability. This test identifies an outlier as one that lies really far away from the central 50% of the data.

> **1.5 × IQR Rule for Outliers**
>
> Any item $x$ in the data set that meets either of the following criteria is suspected of being an outlier:
>
> - $x > Q_3 + 1.5 \times IQR$; or
> - $x < Q_1 - 1.5 \times IQR$

Figure 11.1.: Potential outliers are identified as those that are at least $1.5 \times IQR$ above $Q_3$ or $1.5 \times IQR$ below $Q_1$.

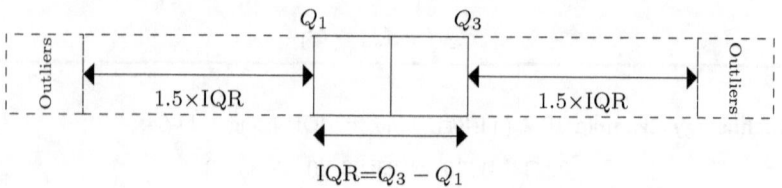

To use the $1.5 \times IQR$ rule:

1. Calculate the $IQR = Q_3 - Q_1$ (ch. 10).

   **Example 11.1.** Identify any possible outlier in the data set from example 10.1. It's IQR has already been calculated as $IQR = 10$.

2. Calculate the number $1.5 \times IQR$

   **(Continuation of example 11.1.)** Since $IQR = 10$, then $1.5 \times IQR = 15$.

3. Calculates $Q_1 - 1.5 \times IQR$ and $Q_3 + 1.5 \times IQR$.

   **(Continuation of example 11.1.)** Since $Q_1 = 41$, $Q_3 = 51$, and $1.5 \times IQR = 15$:

   $$Q_1 - 1.5 \times IQR = 41 - 15 = 26$$
   $$Q_3 + 1.5 \times IQR = 51 + 15 = 66$$

4. If there are any numbers smaller than $Q_1 - 1.5 \times IQR$, classify them as outliers.

   **(Continuation of example 11.1.)** Since $9 < 26 = Q_1 - 1.5 \times IQR$, we classify 9 as an outlier.

5. if there are any numbers larger than $Q_3 + 1.5 \times IQR$, classify them as outliers.

   **(Continuation of example 11.1.)** Since both 76 and 88 are larger than 66, we classify both of them as outliers. Here is an illustration of the calculation:

FAQ 11. What is the IQR Rule for Outliers?

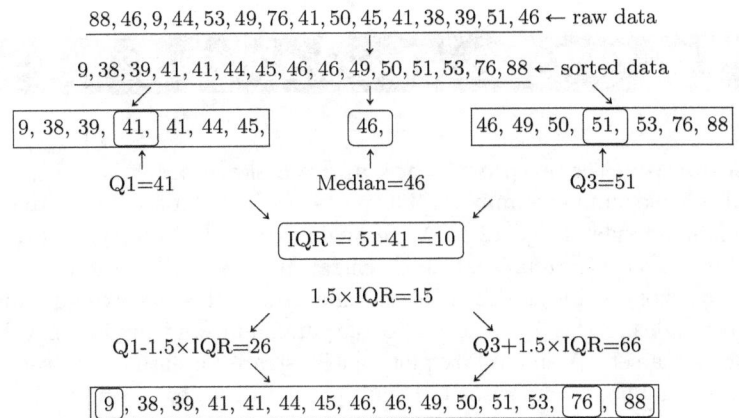

$88, 46, 9, 44, 53, 49, 76, 41, 50, 45, 41, 38, 39, 51, 46 \leftarrow$ raw data

$9, 38, 39, 41, 41, 44, 45, 46, 46, 49, 50, 51, 53, 76, 88 \leftarrow$ sorted data

| 9, 38, 39, [ 41, ] 41, 44, 45, | 46, | 46, 49, 50, [ 51, ] 53, 76, 88 |
|---|---|---|
| Q1=41 | Median=46 | Q3=51 |

$$\text{IQR} = 51\text{-}41 = 10$$

$$1.5 \times \text{IQR} = 15$$

Q1-1.5×IQR=26          Q3+1.5×IQR=66

[ 9 ], 38, 39, 41, 41, 44, 45, 46, 46, 49, 50, 51, 53, [ 76 ], [ 88 ]

# Exercises

For each of the following data data sets, (a) Find the five number summary; (b) Find the IQR; and (c) Determine if there are any outliers according the 1.5×IQR rule.

1. 78, 5, 23, 117, 52, 71, 47, 56, 52, 54, 39, 49, 52, 34, 43, 75, 46 (ans: [5.0, 41.0, 52.0, 63.5, 117.0]; IQR=22.5; outliers=5,117)

2. 35, 45, 53, 57, 66, 12, 61, 49, 44, 44, 45, 47, 55 (ans: [12.0, 44.0, 47.0, 56.0, 66.0]; IQR=12.0; outliers=12)

3. 29, 52, 48, 66, 51, 44, 62, 75 (ans: [29.0, 46.0, 51.5, 64.0, 75.0]; IQR=18.0; outliers=none)

4. 83, 56, 42, 56, 99, 118, 197, 102, 112, 108, 52, 110, 86, 11, 54 (ans: [11.0, 54.0, 86.0, 110.0, 197.0]; IQR=56.0; outliers=197)

5. 1, 6, 8, 10, 7, 13, 13, 11, 8, 10, 9,19 (ans: [1.0, 7.5, 9.5, 12.0, 19.0]; IQR=4.5; outliers=19)

6. 9, 12, 10, 10, 9, 15, 7, 19, 11, 8, 13, 12, 10, 12 (ans: [7.0, 9.0, 10.5, 12.0, 19.0]; IQR=3.0; outliers=19)

7. 10, 11, 11, 13, 14, 11, 8, 12, 11, 7, 11, 13, 7 (ans: [7.0, 9.0, 11.0, 12.5, 14.0]; IQR=3.5; outliers=none)

8. 51.9, 52.3, 52.7, 50.5, 54.7, 51.5, 52.3, 52.7, 53.6 (ans: [50.5, 51.7, 52.3, 53.15, 54.7]; IQR=1.45; outliers=none)

9. 22.0, 23.4, 25.8, 22.2, 22.9, 22.6, 18.2, 22.8, 20.9, 18.0, 18.3, 19.9, 23.0, 21.0, 22.0, 16.8, 19.4, 16.4, 21.1, 21.1, 20.2, 22.2, 17.2, 20.3, 21.5 (ans: [16.4, 18.85, 21.1, 22.4, 25.8]; IQR=3.55; outliers=none)

10. 7428, 7394, 7198, 7122, 7575, 7299, 7169, 7195 (ans: [7122.0, 7182.0, 7248.5, 7411.0, 7575.0]; IQR=229.0; outliers=none)

11. 13.23, 11.74, 12.06, 13.91, 13.76, 14.33, 14.27, 13.96, 13.42, 10.94, 13.04, 13.43, 11.72, 14.44, 13.95, 13.74, 13.22, 13.22, 12.41, 12.85, 14.28, 8.81, 13.53, 12.57, 12.10, 12.77, 14.09, 14.02, 11.30, 13.04, 12.79, 13.87, 14.24, 12.43, 14.18, 13.55, 13.41, 14.37, 13.47, 14.09, 13.91, 13.72, 14.13, 14.13, 17.92, 13.61, 11.37, 12.20, 13.84, 13.77 (ans: [8.81, 12.77, 13.54, 14.02, 17.92]; IQR=1.25; outliers=8.81,17.92)

12. Using the population data set from table 6.1 in chapter 6. (ans: [1.2, 45.8, 103.05, 235.1, 9856.5]; IQR=189.3; outliers=594.8, 738.1, 839.4, 1018.1, 1088.2, 1195.5, 9856.5)

# FAQ 12. How Do I Make a Box Plot?

A **box plot** (or **boxplot**, or **box and whiskers plot**), is a visualization of the five number summary. The parts of a box plot are illustrated on the left hand side of fig. 12.1. A rectangle extending from $Q_1$ to $Q_3$ (along the $y$ direction) represents the central fifty percent of the distribution. A horizontal line is drawn at the median. Whiskers extend from this rectangle, in the $y$ direction, to the maximum and minimum values of the data set. A single box plot can illustrate an entire data set in this way, enabling comparisons of multiple data sets on a single plot (right hand of fig. 12.1).

This is how you make a box plot:

1. **Label the $x$ and $y$ axis.** The $y$ axis should extend from at least the minimum values to the maximum value of the all data sets being plotted. The $x$ axis should have one tick-mark for each data set.

2. **For each data set being plotted:**

   a) Find the **five number summary** (ch. 9).

      **Example 12.1.** Make a box plot of the following data:

      $$88, 46, 9, 44, 53, 49, 76, 41, 50, 45, 41, 38, 39, 51, 46$$

      From example 10.1 the five number summary is [9, 41, 46, 51, 88].

   b) **Identify any outliers** according to the $1.5 \times IQR$ rule (ch. 11).

      **(Continuation of example 12.1.)** From example 10.1, the $IQR$ of this data set is 10; $1.5 \times IQR = 15$; and all points outside the interval [26, 66] are possible outliers. Thus 9, 76, and 88 may be outliers.

   c) Above the $x$-tick for this data set, **draw a rectangle** that extends (in $y$-values) from $Q_1$ to $Q_3$.

      **(Continuation of example 12.1.)** We draw a box extending from $y = 41$ ($Q_1$) to $y = 51$ ($Q_3$). See (fig. 12.2a).

Figure 12.1.: Left: parts of a box plot. Right: A collection box plots comparing multiple data sets. The data points shown beyond the end of the whiskers are suspected outliers.

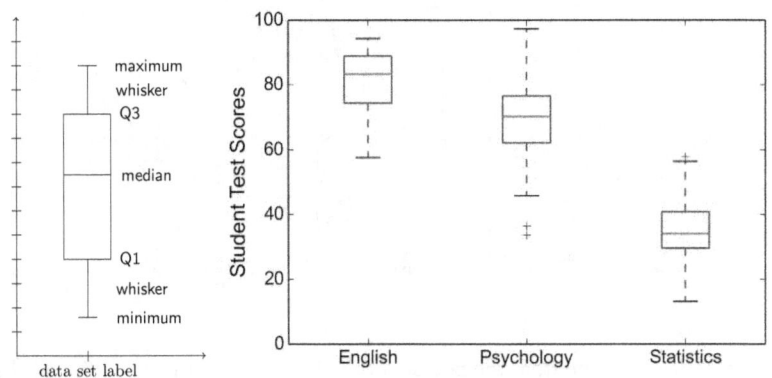

d) Draw a **horizontal line** across the rectangle at the $y$-values corresponding to the **median** of the data set.

(**Continuation of example 12.1.**) We draw a horizontal line across the box at the median, $y = 46$. (see fig 12.2b).

e) **Draw a vertical line** from $Q_3$ to the smaller of $Q_3 + 1.5 \times IQR$ and the maximum of the data set. If there are any outliers larger than $Q_3$, plot them as single points or plus signs.[1]

(**Continuation of example 12.1.**) We draw the top whisker as a vertical line, shown in fig. 12.2c, from the top of the box at $y = 51$ to $y = 66$, and draw in the two top outliers at 76 and 88.

f) **Draw a vertical line** from $Q_1$ to the larger of $Q_1 - 1.5 \times IQR$ and the minimum value of the data. If there are any outliers below $Q_1 - 1.5 \times IQR$, plot them as single points or plus signs.[2]

(**Continuation of example 12.1.**) We draw the bottom whisker as a

---

[1] Rather than extending the whiskers to $Q_1 - 1.5 \times IQR$ on the bottom and $Q_3 + 1.5 \times IQR$ on the top, some authors prefer to extend the whiskers to the 5th percentile on the bottom and the 95th percentile on the top. This will often given similar results for symmetric bell shaped distributions.

[2] See footnote to step 2e.

Figure 12.2.: Making a box plot. Left to right: (a) Draw a box that extends from $Q_1$ to $Q_3$. (b) Draw a line at the median. (c) Draw whisker to maximum. Shown when there are upper outliers. (d) Draw whisker to minimum (shown with lower outliers). (e) Box plot with two boxes. See text.

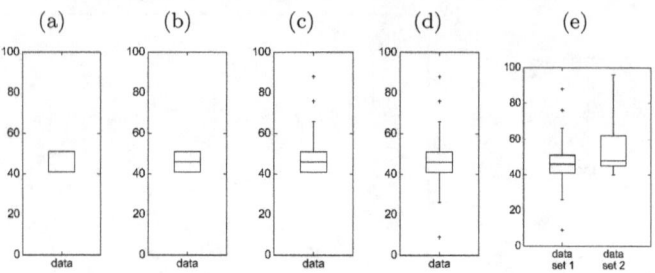

vertical line from the bottom of the box at $y = 41$ to $y = 26$, and then draw a symbol to indicate the outlier at $y = 9$. (fig. 12.2d.)

3. Use one tick-mark on the $x$-axis for each data set, draw one box plot above that tick-mark for each data set.

**Example 12.2.** This is illustrated in fig 12.2e.

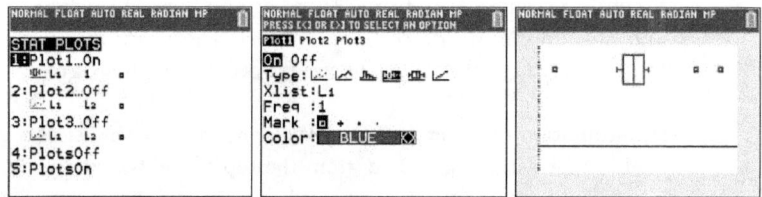

**Technology Example 12.3 (TI-84).** Make a box plot of the data set in example 12.1 using a TI-84.

1. Enter the data into list L1 through the data entry menu, stat ⟩ EDIT ⟩ Edit followed by enter. Navigate to list 1 and with the arrow keys and type each number followed by enter (e.g., see example 6.2).
2. Hit 2nd ⟩ stat plot ⟩ 1 then enter (left figure below).
3. When the plot1 menu appears, navigate to On with the arrow keys and select with enter (right figure below).
4. Then on the same menu, on the Type line, select the box plot (fourth icon).

FAQ 12. How Do I Make a Box Plot?

5. Verify `Xlist` is `L1`.
6. Hit `graph`. If necessary, hit `zoom` to change the scale.

**Technology Example 12.4 (R).** Make a box plot of the the data set in example 12.1 using R.

Do this with the function **boxplot**.

```
> x=c(88,46,9,44,53,49,76,41,50,45,41,38,39,51,46)
> boxplot(x)
```

The plot will pop up in a separate window; in RStudio, in will appear in the plot window.

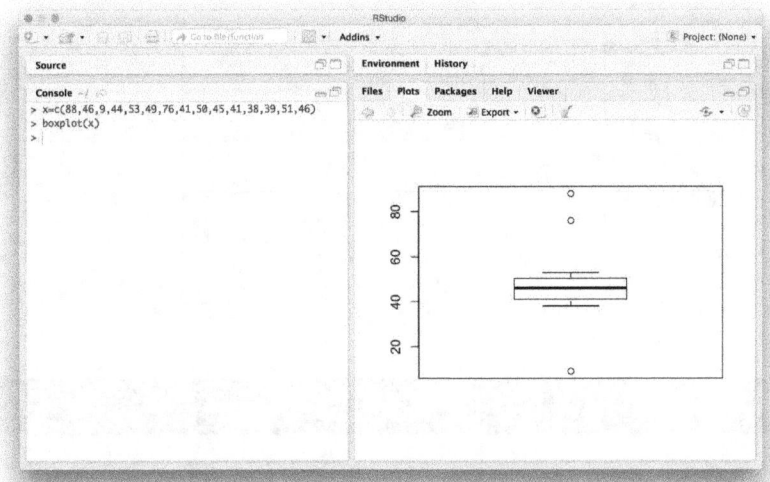

# Exercises

Make box plots for each of the data sets listed in the exercises in ch. 9.

1. 78, 48, 52, 71, 47, 56, 52, 54, 39, 49, 52, 34, 43, 75, 46.
2. 35, 45, 53, 57, 66, 61, 49, 44, 44, 45, 47, 55
3. 29, 52, 48, 66, 51, 44, 62
4. 83, 56, 42, 56, 99, 118, 147, 102, 112, 108, 52, 110, 86, 50, 54

5. 6, 8, 10, 7, 13, 13, 11, 8, 10, 9
6. 9, 12, 10, 10, 9, 15, 7, 11, 8, 13, 12, 10, 12
7. 10, 11, 11, 13, 11, 8, 12, 11, 7, 11, 13, 7
8. 51.9, 52.3, 52.7, 50.5, 51.5, 52.3, 52.7, 53.6
9. 22.0, 23.4, 25.8, 22.2, 22.9, 22.6, 18.2, 22.8, 20.9, 18.0, 18.3, 19.9, 23.0, 21.0, 22.0, 16.8, 19.4, 16.4, 21.1, 21.1, 20.2, 22.2, 17.2, 20.3, 21.5
10. 7428, 7394, 7198, 7122, 7575, 7299,

7169, 7195

11. 13.23, 11.74, 12.06, 13.91, 13.76,
14.33, 14.27, 13.96, 13.42, 10.94,
13.04, 13.43, 11.72, 14.44, 13.95,
13.74, 13.22, 13.22, 12.41, 12.85,
14.28, 12.81, 13.53, 12.57, 12.10,

12.77, 14.09, 14.02, 11.30, 13.04,
12.79, 13.87, 14.24, 12.43, 14.18,
13.55, 13.41, 14.37, 13.47, 14.09,
13.91, 13.72, 14.13, 14.13, 13.92,
13.61, 11.37, 12.20, 13.84, 13.77

Answers:

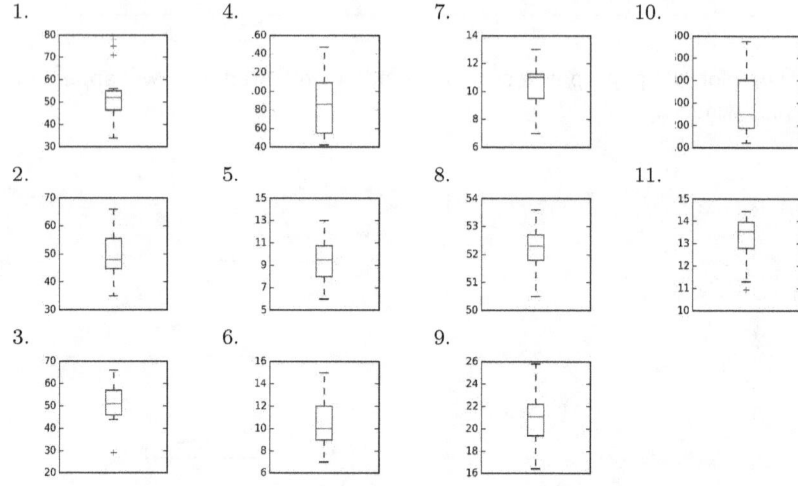

1.

4.

7.

10.

2.

5.

8.

11.

3.

6.

9.

# FAQ 13. How Do I Find the Mean?

The **mean** is the average of all the numbers in a data set. To find
the mean, add up all the numbers and divide by the total count. In a
continuous distribution, the mean is the balance point.

## Sample Mean

The **mean**, or **sample mean**, of $x_1, x_2, \ldots, x_n$ is the average

$$\overline{x} = \frac{1}{n} \sum_{i=1}^{n} x_i = \frac{x_1 + x_2 + \cdots + x_n}{n}$$

We read the number $\bar{x}$ as **x bar**

We read the symbol $\displaystyle\sum_{i=1}^{n}$ as "The sum from $i$ equals 1 to $i$ equals $n$ of ..." (i.e., whatever is written next to the upper-case Greek Sigma $\Sigma$).

The shorthand notation $\sum x_i$ is frequently used without the indices. It means "the sum of all of the $x$-values." When we write $\bar{x} = \dfrac{1}{n}\sum x_i$ we mean sum over all the $x_i$'s and then divide by $n$.

To find the mean of a data set:

1. **Count** the number of items in the data set. Call this number $n$.

   **Example 13.1.** Find the mean of following data set:

   $$22, 33, 31, 36, 23, 25, 23, 20, 35, 33, 33$$

   There are $n = 11$ items in this data set.

2. **Add** up all of the values in the data set. Call this sum $\sum x_i$.

   **(Continuation of example 13.1.)**

   $$\sum x_i = 22 + 33 + 31 + 36 + 23 + 25 + 23 + 20 + 35 + 33 + 33 = 314$$

3. **Calculate the mean** as the sum (from step 2) divided by the count (from step 1).

   $$\bar{x} = \frac{\sum x_i}{n}$$

   You can calculate the mean in step 3 either as the quotient

   $$\bar{x} = \frac{\sum x_i}{n}$$

   or as the product

   $$\bar{x} = \left(\frac{1}{n}\right) \times \left(\sum x_i\right)$$

   as they are mathematically equivalent formulas.

---

(**Continuation of example 13.1.**) The mean is

$$\bar{x} = \frac{22 + 33 + 31 + 36 + 23 + 25 + 23 + 20 + 35 + 33 + 33}{11}$$

$$= \frac{314}{11}$$

$$= 28.545$$

**Note:** In general, the mean will be different from the median. If the data is skewed, the mean tends to be pulled away from the median and towards the long tail.

**Technology Example 13.2 (TI-84).** Find the mean using a TI-84:

$$22, 33, 31, 36, 23, 25, 23, 20, 35, 33, 33$$

1. Select [stat] [EDIT] ⟩ [Edit] [enter]. This takes you to the data entry page.
2. Navigate to the L1 column using the arrow keys.
3. Enter the numbers in the column L1, following each number by the [enter] (e.g., see example 6.2). When you are done select [2nd] [quit].
4. Select [2nd] [list] then [MATH] ⟩ [mean] [enter].
5. Select [2nd] [L1] [ ) ] [enter]. The mean will be displayed on the screen.
6. Alternatively, the mean can be calculated as part of one-variable statistics. Type [stat] [CALC] ⟩ [1-Var Stats] [enter] to get to the 1-variable statistics menu. On the first line select [List] ⟩ [L1]. This selects the data in list L1. Navigate to [Calculate:] and press [enter]. The mean is displayed on the line labeled $\bar{x}$. The remainder of the items displayed are summarized in table 7.1.

**Technology Example 13.3 (R).** Find the mean using R:

$$22, 33, 31, 36, 23, 25, 23, 20, 35, 33, 33$$

Use the function **mean**:

```
> x=c(22,33,31,36,23,25,23,20,35,33,33)
> mean(x)
[1] 28.54545
```

**Technology Example 13.4 (Spreadsheet).** Find the mean of the following data using a spreadsheet:

$$22, 33, 31, 36, 23, 25, 23, 20, 35, 33, 33$$

1. Enter the numbers in a column in the spreadsheet.
2. In any cell type the expression [**=average(**].

3. Place your cursor just to the right of the right parenthesis, then use the mouse to select the data items you want in the column of data. The addresses of the data items will appear.

4. Type a ⎡ ) ⎤ and then ⎡enter⎤. The answer will appear in the same cell.

| =average(D8:D18) |
|---|
| 22 |
| 33 |
| 31 |
| 36 |
| 23 |
| 25 |
| 23 |
| 20 |
| 35 |
| 33 |
| 33 |

| 28.5454545 |
|---|
| 22 |
| 33 |
| 31 |
| 36 |
| 23 |
| 25 |
| 23 |
| 20 |
| 35 |
| 33 |
| 33 |

# Exercises

Find the mean of each of the following lists of numbers. Compare with the median of the data.

1. 78, 48, 52, 71, 47, 56, 52, 54, 39, 49, 52, 34, 43, 75, 46 (ans: median=52.0; mean=53.07)

2. 35, 45, 53, 57, 66, 61, 49, 44, 44, 45, 47, 55 (ans: median=48.0; mean=50.08)

3. 29, 52, 48, 66, 51, 44, 62 (ans: median=51.0; mean=50.29)

4. 83, 56, 42, 56, 99, 118, 147, 102, 112, 108, 52, 110, 86, 50, 54 (ans: median=86.0; mean=85.0)

5. 6, 8, 10, 7, 13, 13, 11, 8, 10, 9 (ans: median=9.5; mean=9.5)

6. 9, 12, 10, 10, 9, 15, 7, 11, 8, 13, 12, 10, 12 (ans: median=10.0; mean=10.62)

7. 10, 11, 11, 13, 11, 8, 12, 11, 7, 11, 13, 7 (ans: median=10.0; mean=10.62)

8. 42.2, 49.6, 43.3, 64.8, 53.7, 42.3, 55.5 (ans: median=49.6; mean=50.2)

9. 73.7, 71.4, 70.8, 72.2, 73.4, 74.0, 72.3 (ans: median=72.2; mean=72.5)

10. 55.9, 54.4, 52.9, 47.6, 44.2, 52.2, 54.3, 53.2 (ans: median=53.0; mean=51.84)

11. 22.0, 23.4, 25.8, 22.2, 22.9, 22.6, 18.2, 22.8, 20.9, 18.0, 18.3, 19.9, 23.0, 21.0, 22.0, 16.8, 19.4, 16.4, 21.1, 21.1, 20.2, 22.2, 17.2, 20.3, 21.5 (ans: median=21.1; mean=20.77)

12. 13.23, 11.74, 12.06, 13.91, 13.76, 14.33, 14.27, 13.96, 13.42, 10.94, 13.04, 13.43, 11.72, 14.44, 13.95, 13.74, 13.22, 13.22, 12.41, 12.85, 14.28, 12.81, 13.53, 12.57, 12.10, 12.77, 14.09, 14.02, 11.30, 13.04, 12.79, 13.87, 14.24, 12.43, 14.18, 13.55, 13.41, 14.37, 13.47, 14.09, 13.91, 13.72, 14.13, 14.13, 13.92, 13.61, 11.37, 12.20, 13.84, 13.77 (ans: median=13.5; mean=13.3)

13. Using the population data set from table 6.1 in chapter 6. (ans: median=103.1; mean=397.94)

# FAQ 14. How Do I Find the Variance?

The **variance** is a measure of the spread of a distribution. It depends on the mean (ch. 13), and is the square of the standard deviation (ch. 15).

Calculation of the variance is fairly messy and the full calculation (which is explained here) is rarely done manually because it is so error prone. Statistical software or calculators are generally used instead. Software is preferable to calculators, since, unless your calculator has the option to display the entire data list, it is generally difficult to go back verify that your data has been correctly entered. There are also many websites and free cell phone apps that will calculate the variance.

---

### Sample Variance

The **variance** $s^2$ of a sample of size $n$, $\{x_1, x_2, \ldots, x_n\}$, is

$$s^2 = \frac{1}{n-1} \sum_{i=1}^{n} (x_i - \bar{x})^2 = \frac{(x_1 - \bar{x})^2 + (x_2 - \bar{x})^2 + \cdots + (x_n - \bar{x})^2}{n-1}$$

where $\bar{x}$ is the sample mean $\bar{x} = \dfrac{1}{n} \sum_{i=1}^{n} x_i$ (ch. 13).

---

To calculate the variance:

1. **Find the mean, $\bar{x}$, of the data** (ch. 13).

   **Example 14.1.** Find the variance of the following data set:

   $$22, 33, 31, 36, 23, 25, 23, 20, 35, 33, 33$$

   $$\bar{x} = \frac{22 + 33 + 31 + 36 + 23 + 25 + 23 + 20 + 35 + 33 + 33}{11}$$
   $$= \frac{314}{11}$$
   $$= 28.545$$

2. **Write out the expression for the sum $\sum$:**

$$\sum = \sum_{i=1}^{n}(x_i - \bar{x})^2$$

accounting for all $n$ terms.

(**Continuation of example 14.1.**) Since there are $n = 11$ terms,

$$\begin{aligned}\sum = {}& (x_1 - \bar{x})^2 + (x_2 - \bar{x})^2 + (x_3 - \bar{x})^2 + \\ & + (x_4 - \bar{x})^2 + (x_5 - \bar{x})^2 + (x_6 - \bar{x})^2 \\ & + (x_7 - \bar{x})^2 + (x_8 - \bar{x})^2 + (x_9 - \bar{x})^2 \\ & + (x_{10} - \bar{x})^2 + (x_{11} - \bar{x})^2\end{aligned}$$

3. Everywhere in the expression for $\sum$, **replace the value of $\bar{x}$ with the number calculated in step 1.**

(**Continuation of example 14.1.**) In step 1 we found $\bar{x} = 28.545$;

$$\begin{aligned}\sum = {}& (x_1 - 28.545)^2 + (x_2 - 28.545)^2 + (x_3 - 28.545)^2 + \\ & + (x_4 - 28.545)^2 + (x_5 - 28.545)^2 + (x_6 - 28.545)^2 \\ & + (x_7 - 28.545)^2 + (x_8 - 28.545)^2 + (x_9 - 28.545)^2 \\ & + (x_{10} - 28.545)^2 + (x_{11} - 28.545)^2\end{aligned}$$

4. **Fill in the actual values of the $x_1, \ldots, x_n$** from the data into the equation for $\sum$.

(**Continuation of example 14.1.**) Using the data given, we fill in $x_1 = 22$, $x_2 = 33$, $x_3 = 31$, and so forth.

$$\begin{aligned}\sum = {}& (22 - 28.545)^2 + (33 - 28.545)^2 + (31 - 28.545)^2 \\ & + (36 - 28.545)^2 + (23 - 28.545)^2 + (25 - 28.545)^2 \\ & + (23 - 28.545)^2 + (20 - 28.545)^2 + (35 - 28.545)^2 \\ & + (33 - 28.545)^2 + (33 - 28.545)^2\end{aligned}$$

5. Perform the **subtractions** in each term.

(**Continuation of example 14.1.**)

$$\begin{aligned}\sum = {}& (-6.545)^2 + 4.455^2 + 2.455^2 \\ & + 7.455^2 + (-5.545)^2 + (-3.545)^2 + (-5.545)^2 \\ & + (-8.545)^2 + 6.455^2 + 4.455^2 + 4.455^2\end{aligned}$$

6. **Square all the indicated terms** in parenthesis, and add the resulting sum.

(**Continuation of example 14.1.**)

$$\sum = 42.837 + 19.847 + 6.027 + 55.577 + 30.747 + 12.567$$
$$+ 30.747 + 73.017 + 41.667 + 19.847 + 19.847$$
$$= 352.727$$

7. **Substitute** the result for $\sum$ into the equation of the variance.

(**Continuation of example 14.1.**) There are $n = 11$ data values, and therefore the sample variance is

$$s^2 = \frac{1}{n-1} \sum_{i=1}^{n} (x_i - \bar{x})^2 = \frac{1}{11-1} \times \sum = \frac{352.727}{10} = 35.27$$

**Technology Example 14.2 (TI-84).** Find the variance using a TI-84:

$$22, 33, 31, 36, 23, 25, 23, 20, 35, 33, 33$$

1. Select [stat] [EDIT ⟩ Edit] [enter]. This takes you to the data entry page.
2. Navigate to the L1 column using the arrow keys.
3. Enter the numbers in the column L1, following each number by the [enter] (see, e.g. example 8.5). When you are done type [2nd] [quit].
4. Select [2nd] [list] then [MATH ⟩ variance] [enter].
5. Select [2nd] [L1] [ ) ] [enter]. The variance will be displayed on the screen.
6. Alternatively, at step 4, type [stat] [CALC ⟩ 1-Var Stats] [enter] to go to the 1-variable statistics menu. On the first line select [List ⟩ L1]. This selects the data in list L1. Navigate to [Calculate:] and press [enter]. The variance is square of the standard deviation, which is labeled as Sx.

**Technology Example 14.3 (R).** Find the variance using R:

$$22, 33, 31, 36, 23, 25, 23, 20, 35, 33, 33$$

Use the function **var**:

```
> x=c(22,33,31,36,23,25,23,20,35,33,33)
> var(x)
[1] 35.27273
```

**Technology Example 14.4 (Spreadsheet).** Find the variance of the following data using a spreadsheet:

$$22, 33, 31, 36, 23, 25, 23, 20, 35, 33, 33$$

1. Enter the numbers in a column in the spreadsheet.
2. In any cell type the expression `=var(`.
3. Place your cursor just to the right of the right parenthesis, then use the mouse to select the data items you want in the column of data. The addresses of the data items will appear.
4. Type a `)` and then `enter`. The answer will appear in the cell.

| =var(D8:D18) | | 35.2727273 |
|---|---|---|
| 22 | | 22 |
| 33 | | 33 |
| 31 | | 31 |
| 36 | | 36 |
| 23 | | 23 |
| 25 | | 25 |
| 23 | | 23 |
| 20 | | 20 |
| 35 | | 35 |
| 33 | | 33 |
| 33 | | 33 |

# Exercises

Find the variances of each of the following data sets.

1. 78, 48, 52, 71, 47, 56, 52, 54, 39, 49, 52, 34, 43, 75, 46 (ans: 159.21; mean=53.07)
2. 35, 45, 53, 57, 66, 61, 49, 44, 44, 45, 47, 55 (ans: 74.27; mean=50.08 )
3. 29, 52, 48, 66, 51, 44, 62 (ans: 147.57; mean=50.29)
4. 83, 56, 42, 56, 99, 118, 147, 102, 112, 108, 52, 110, 86, 50, 54 (ans: 1008; mean=85.0)
5. 6, 8, 10, 7, 13, 13, 11, 8, 10, 9 (ans: 5.61; mean=9.5)
6. 9, 12, 10, 10, 9, 15, 7, 11, 8, 13, 12, 10, 12 (ans: 4.76; mean=10.62)
7. 10, 11, 11, 13, 11, 8, 12, 11, 7, 11, 13, 7 (ans: 4.27; mean=10.62)
8. 42.2, 49.6, 43.3, 64.8, 53.7, 42.3, 55.5 (ans: 71.3; mean=50.2)

9. 73.7, 71.4, 70.8, 72.2, 73.4, 74.0, 72.3 (ans: 1.45; mean=72.5)
10. 55.9, 54.4, 52.9, 47.6, 44.2, 52.2, 54.3, 53.2 (ans: 15.51; mean=51.84)
11. 22.0, 23.4, 25.8, 22.2, 22.9, 22.6, 18.2, 22.8, 20.9, 18.0, 18.3, 19.9, 23.0, 21.0, 22.0, 16.8, 19.4, 16.4, 21.1, 21.1, 20.2, 22.2, 17.2, 20.3, 21.5 (ans: 5.35; mean=20.77)
12. 13.23, 11.74, 12.06, 13.91, 13.76, 14.33, 14.27, 13.96, 13.42, 10.94, 13.04, 13.43, 11.72, 14.44, 13.95, 13.74, 13.22, 13.22, 12.41, 12.85, 14.28, 12.81, 13.53, 12.57, 12.10, 12.77, 14.09, 14.02, 11.30, 13.04, 12.79, 13.87, 14.24, 12.43, 14.18, 13.55, 13.41, 14.37, 13.47, 14.09, 13.91, 13.72, 14.13, 14.13, 13.92, 13.61, 11.37, 12.20, 13.84, 13.77 (ans: 0.81; mean=13.3)
13. Using the population data set from table 6.1 in chapter 6. (ans: 1,869,433; mean=397.94)

# FAQ 15. How Do I Find the Standard Deviation?

The **standard deviation** is probably the most commonly quoted measure of the spread of a distribution. It depends on the mean (ch. 13), and it is the square root of the variance (ch. 14).

---

### Sample Standard Deviation

The **standard deviation** $s$ of a sample of size $n$, $\{x_1, x_2, \ldots, x_n\}$ is

$$s = \sqrt{\frac{\sum_{i=1}^{n}(x_i - \bar{x})^2}{n-1}} = \sqrt{\frac{(x_1 - \bar{x})^2 + (x_2 - \bar{x})^2 + \cdots + (x_n - \bar{x})^2}{n-1}}$$

where $\bar{x}$ is the sample mean $\bar{x} = \dfrac{1}{n}\sum_{i=1}^{n} x_i$ (chapter 13).

---

Calculation of the standard deviation is fairly messy and the full calculation is rarely done by hand in practice because it is so error prone. Statistical software or calculators are generally used instead. Software is preferable to calculators, since, unless your calculator has the option to display the entire data list (for verification), it is difficult to verify that your data has been correctly entered. There are also many websites and free cell phone apps that will calculate the standard deviation.

To calculate the standard deviation:

1. Calculate the **variance** $s^2$ (ch. 14).

   **Example 15.1.** Find the standard deviation:

   $$22, 33, 31, 36, 23, 25, 23, 20, 35, 33, 33$$

   In example 14.1 we determined that $s^2 = 35.273$.

2. Calculate the **square root** of the variance. This is the standard deviation.

---

(**Continuation of example 15.1.**) Since $s^2 = 35.273$,

$$s = \sqrt{s^2} = \sqrt{35.273} = 5.939$$

the standard deviation of the data set is $s = 5.939$

**Technology Example 15.2 (TI-84).** Find the standard deviation using a TI-84:

$$22, 33, 31, 36, 23, 25, 23, 20, 35, 33, 33$$

1. Select stat EDIT ⟩⟩ Edit enter. This takes you to the data entry page.

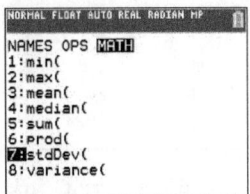 

2. Navigate to the L1 column using the arrow keys.
3. Enter the numbers in the column L1, following each number by the enter (see, e.g. example 8.5). When you are done type 2nd quit.
4. Select 2nd list then MATH ⟩⟩ stdDev enter.
5. Select 2nd L1 ) enter. The standard deviation 5.939084717 will be displayed on the screen.
6. Alternatively, at step 4, type stat CALC ⟩⟩ 1-Var Stats enter to get to the 1-variable statistics menu. On the first line select List ⟩⟩ L1. This selects the data in list L1. Navigate to Calculate: and press enter. The sample standard deviation will be displayed on the line labeled Sx. (See the output screen displayed in example 8.5.) In this case, we can read off the number next to Sx which tells us that $s = 5.939084717$.

**Technology Example 15.3 (R).** Find the standard deviation using R:

$$22, 33, 31, 36, 23, 25, 23, 20, 35, 33, 33$$

Use the function **sd**:

```
> x=c(22,33,31,36,23,25,23,20,35,33,33)
> sd(x)
[1] 5.939085
```

**Technology Example 15.4 (Spreadsheet).** Find the standard deviation of the following data using a spreadsheet:

$$22, 33, 31, 36, 23, 25, 23, 20, 35, 33, 33$$

1. Enter the numbers in a column in the spreadsheet.
2. In any cell type the expression `=stdev(`.
3. Place your cursor just to the right of the right parenthesis, then use the mouse to select the data items you want in the column of data. The addresses of the data items will appear.
4. Type a `)` and then `enter`. The answer will appear in the cell.

| =stdev(D8:D18) | | 5.93908472 |
| --- | --- | --- |
| 22 | | 22 |
| 33 | | 33 |
| 31 | | 31 |
| 36 | | 36 |
| 23 | | 23 |
| 25 | | 25 |
| 23 | | 23 |
| 20 | | 20 |
| 35 | | 35 |
| 33 | | 33 |
| 33 | | 33 |

# Exercises

Find the standard deviations of each of the following data sets.

1. 78, 48, 52, 71, 47, 56, 52, 54, 39, 49, 52, 34, 43, 75, 46 (ans: 12.62; mean=53.07)
2. 35, 45, 53, 57, 66, 61, 49, 44, 44, 45, 47, 55 (ans: 8.62; mean=50.08)
3. 29, 52, 48, 66, 51, 44, 62 (ans: 12.1; mean=7.09)
4. 83, 56, 42, 56, 99, 118, 147, 102, 112, 108, 52, 110, 86, 50, 54 (ans: 31.7; mean=85.0)
5. 6, 8, 10, 7, 13, 13, 11, 8, 10, 9 (ans: 5.61; mean=9.5)
6. 9, 12, 10, 10, 9, 15, 7, 11, 8, 13, 12, 10, 12 (ans: 2.18; mean=10.62)
7. 10, 11, 11, 13, 11, 8, 12, 11, 7, 11, 13, 7 (ans: 2.06; mean=10.62)
8. 42.2, 49.6, 43.3, 64.8, 53.7, 42.3, 55.5 (ans: 8.44; mean=50.2)

9. 73.7, 71.4, 70.8, 72.2, 73.4, 74.0, 72.3 (ans: 1.20; mean=72.5)
10. 55.9, 54.4, 52.9, 47.6, 44.2, 52.2, 54.3, 53.2 (ans: 3.94; mean=51.84)
11. 22.0, 23.4, 25.8, 22.2, 22.9, 22.6, 18.2, 22.8, 20.9, 18.0, 18.3, 19.9, 23.0, 21.0, 22.0, 16.8, 19.4, 16.4, 21.1, 21.1, 20.2, 22.2, 17.2, 20.3, 21.5 (ans: 2.31; mean=20.77)
12. 13.23, 11.74, 12.06, 13.91, 13.76, 14.33, 14.27, 13.96, 13.42, 10.94, 13.04, 13.43, 11.72, 14.44, 13.95, 13.74, 13.22, 13.22, 12.41, 12.85, 14.28, 12.81, 13.53, 12.57, 12.10, 12.77, 14.09, 14.02, 11.30, 13.04, 12.79, 13.87, 14.24, 12.43, 14.18, 13.55, 13.41, 14.37, 13.47, 14.09, 13.91, 13.72, 14.13, 14.13, 13.92, 13.61, 11.37, 12.20, 13.84, 13.77 (ans: 0.9; mean=13.3)
13. Using the population data set from table 6.1 in chapter 6. (ans: 1.367; mean=397.94)

# FAQ 16. What is a Density Function? What is the Difference Between Distribution and Density?

The **distribution** of population or a data set is a description of the actual data or population. It describes:

- ▸ The actual values of the data; and
- ▸ How frequently each value occurs.

Distributions are usually illustrated with. Categorical variables may be described by **bar graphs** or **pie charts**. Quantitative variables may be described by **histograms** (ch. 6) and **stem plots** (ch. 5).

Two types of density and distribution functions are used in statistics making things more confusing:

- ▸ The **PDF** or **Probability Density Function**
- ▸ The **CDF** or **Cumulative Distribution Function**

The **PDF (Probability Density Function)** $f(x)$ is an idealized model of the histogram. It is a smooth curve that looks like the histogram but is re-scaled so that the total area under the curve is equal to one (fig. 16.1). A PDF has the following properties:

- ▸ The values of the PDF function are always positive.
- ▸ The total area under the curve, between the PDF and the $x$-axis, is equal to one.
- ▸ The area under the curve, between the PDF and the $x$-axis, and between any two points $x = a$ and $x = b$, where $a < b$, is the proportion of the population or the data set that has values between $a$ and $b$ (fig. 16.2).

The value of the PDF is never actually needed in the calculation of any statistical quantities in basic statistics. This is fortunate because the formulas for many of the PDF's that are actually used most commonly (such as the normal) are very complicated.

Figure 16.1.: The PDF is an idealized approximation of the histogram. In this figure, the PDF has been rescaled to show how it might closely match the histogram of a large sample or population.

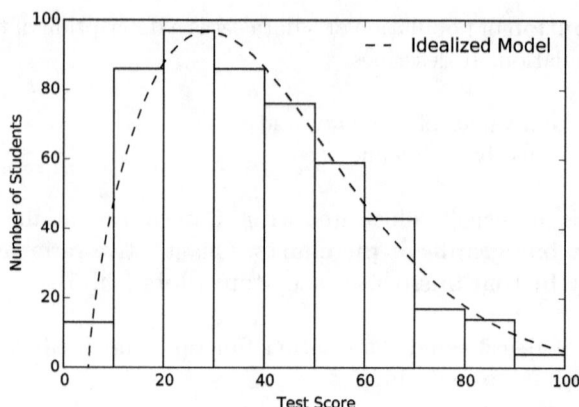

Figure 16.2.: The area of the region bounded by the lines $x = a$, $x = b$, the PDF, and the $x$ axis gives the proportion of the population with $x$ values between $a$ and $b$.

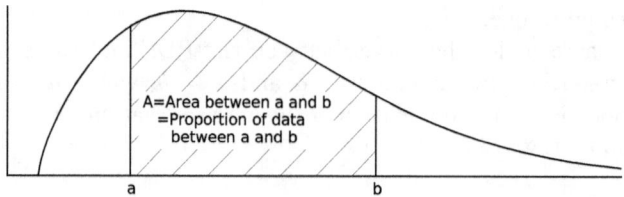

Figure 16.3.: The CDF is defined as the area (to the left of $x$) under the curve of the PDF.

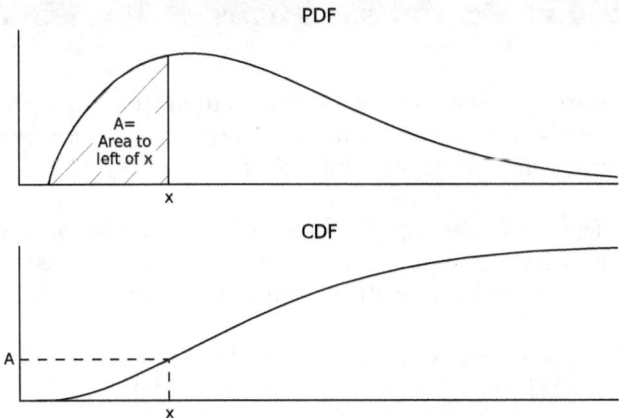

The **CDF** or **Cumulative Distribution Function** $F(x)$, may be defined in terms of the PDF. The value of $F(a)$ is the area $A$ bounded by the line $x = a$, the curve of $f(x)$ (the PDF), and the $x$-axis (fig. 16.3). A CDF has the following properties:

- It is always positive.
- It is an increasing function of $x$.
- For extremely large values of $x$, it becomes very close to one.
- For extremely large and negative values of $x$, it becomes very close to zero.

The numbers in table A.1 are the values of the CDF of a special distribution called the normal distribution. We will discuss how to read tables like table A.1 in ch. 21. We will also use the CDF to calculate proportions and probabilities of the distribution (ch. 23 – 26). The reason why these values are listed in tables rather than just given as a formula for the normal distribution is because no such formula exists.

# FAQ 17. What are Shape, Center, and Variability?

**Shape, center** and **variability** are general terms that we use to describe the features of a distribution. Some of the features may only be described qualitatively, while others are described quantitatively.

**Shape** refers to the overall shape of a distribution. The most commonly used distributions in introductory statistics are bell-shaped (i.e., they look like a cross section of bell). Common shapes are (fig. 17.1):

- bell shaped such as the normal (ch. 18) and t distribution (ch. 37)
- skewed bell shaped such as $\chi^2$ distribution (ch. 56)
- triangular shaped distributions
- uniform (constant) distribution

Shape may also refer to other properties of the distribution such as:

- **Skewness** - are there more points on one side of the maximum than on the other? If there is a long tail on one side, the distribution is said to be skewed in the direction of the tail (**skewed left** or **skewed right**).
- Number of peaks or **modes** - is there are single maximum, or multiple peaks and valleys? A uniform distribution has no peaks; bell shaped and triangular distributions have a single peak. If there are multiple peaks, the distribution is called **multimodal**.
- Is the distribution **symmetric**? Is there some value of $x$ where you can draw a vertical line, then fold over a picture of the distri-

Figure 17.1.: Various types of density functions are shown here (not to scale).

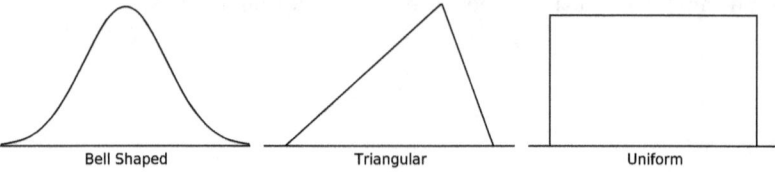

| Bell Shaped | Triangular | Uniform |

Figure 17.2.: Illustration of skewed left and right, symmetric, and multimodal distributions

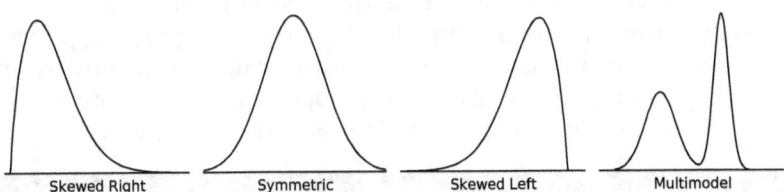

bution so that the right-half falls on top of the left-half? Normal distributions and t-distributions are symmetric.

▸ Are there any **holes** in the data? Do any data points deviate deviate from the general pattern set by the rest of the points? We call these unusual points **outliers**. Outliers are usually separated from the rest of the data by large holes or missing data, although holes do not necessarily indicate outliers.

**Center** is usually quantified in one of two ways:

▸ The **mean** or **average** (ch. 13) is the balancing of a data set.
▸ The **median** (ch. 7) is the value of $x$ at which half of the data is smaller than $x$ and half of the data is larger than $x$.

The mean and median are rarely equal, except in symmetric distributions. In skewed distributions, the mean is always pulled further out to the long tail of skewness than the median.

Another measure of center is the **mode**, although the mode does not always come anywhere near either the median or mean. The mode is the maximum value of a distribution. If it is multimodal (if there are multiple peaks) then each of these are called modes.

**Variability** is a general term for describing the **spread** or **width** of a distribution. The following measures are used to described variability:

▸ The **maximum** and **minimum** values of $x$
▸ The **Inter-Quartile Range** (IQR) gives the the values that bound the central 50% of the data (ch. 10).
▸ The **Five Number Summary** combines lists the quartiles. These are the boundaries between each quarter of the data, and include

the maximum and minimum (ch. 9).

- ▸ The **variance** measures average of the sum of the squares of the differences of the $x$ values from the mean of the data (ch. 14).
- ▸ The **standard deviation** (ch. 15) is the square root of the variance. It is not itself a measure of the width, but related to the width. In a normal distribution, approximately 99% of the data falls within three standard deviations of the mean (ch. 20).

## FAQ 18. What is a Normal Distribution? What is the Difference Between a Normal Distribution and a Standard Normal Distribution?

A **normal distribution**, also called a **Gaussian distribution**, is one of the most commonly used distributions in statistics. It is useful because of the central limit theorem (ch. 28), which tells us that under the right conditions (which in fact never really occur) nearly all things can be reasonably well approximated by a normal distribution. Many scientists incorrectly believe that if you take a large enough sample than your data will approach a normal distribution, but in fact, this is not true. If the population has the right distribution properties, and you take infinitely many samples, and those samples are properly chosen, and you find the means of those samples, then distributions of those means will approach a normal distribution. It is this purely abstract property of the normal distribution which allows all of elementary inferential statistics to work.

Normal distributions are described by two parameters: the mean $\mu$ and standard deviation $\sigma$. The mathematical equation for a normal density function is not useful in elementary statistics so we will not present it here. The density function is bell shaped and goes nearly to zero within a little over 3 standard deviations of the mean. The variability and shape of a normal distribution is summarized by the 68-95-99.7 rule (ch. 20). The values of the PDF and CDF can be determined directly on calculators, e.g., with [2nd] [distr] **normaldf** or [2nd] [distr] **normalcdf** on the TI-84.

It is common in statistical problems to perform a **normalization**. This means converting values from a normal distribution to a **standardized** scale. A **standard normal distribution** has a mean of zero and a standard deviation of one. The conversion between a normal distribution and its standard normal equivalent is called a **z-score** (ch. 19). Normalized values are tabulated for easy calculation (table A.1).

# FAQ 19. What is z?

The **z-value** or **z-score** measures how far a data value falls from the mean in units of the standard deviation. It provides a normalization that allows us to compare values from different populations, for example, when both populations are known to be distributed normally but with different means and standard deviation.

The **z-score** of a variable is only meaningful when the variable has a normal distribution.

Figure 19.1.: Conversion between raw data $x$ and $z$ score.

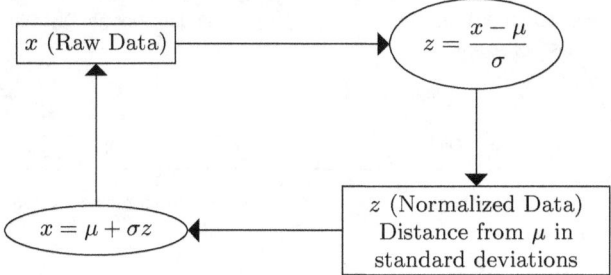

You must know the population mean ($\mu$) and standard deviation ($\sigma$). The $z$-score measures the corresponding $x$ data value as its distance in standard deviations from the mean. You can convert back and forth as illustrated fig. 19.1.

**Finding $z$ given $x$:**   If $x$ is selected from a population that is normally distributed with mean $\mu$ and standard deviation $\sigma$, then

$$z = \frac{x - \mu}{\sigma}$$

**Example 19.1.** The height of American men ages 20 to 29 is distributed approximately normally with mean $\mu = 69.4$ inches with standard deviation $\sigma = 3.9$

---

inches. Dick is $5' - 11''$ or 71 inches tall. The the z-value for his height is

$$z = \frac{71 - 69.4}{3.9} = \frac{1.6}{3.9} = 0.41$$

The average height of American women in the same age group is $\mu = 64.2$ inches with standard deviation $\sigma = 2.8$. Jane is $5' - 7''$ or 67 inches tall. To she if she is taller shorter than Dick, measured with respect to her population, we can calculate her height

$$z = \frac{67 - 64.2}{2.8} = \frac{2.8}{2.8} = 1.0$$

Dick's height, at 71 inches, is $0.41\sigma$ above the mean; Jane's height, at 67 inches, is $\sigma$ above the mean. So Jane is taller, even though she is four inches shorter (fig. 19.2).

Figure 19.2.: Left: Jane is tall for a woman; Dick is just above average for a man. When you normalize both scores (right), even though Jane's actual height in inches is less than that of Dick's, her normalized hight is greater than that of Dick's.

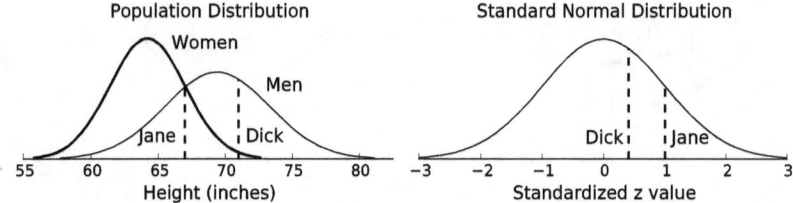

**Finding $x$ given a z-value:** Use the following formula: $x = \mu + z\sigma$.

**Example 19.2.** The heights of American men and women aged $20 - 29$ are distributed normally with means of 69.4 inches (men) and 64.2 inches (women), and standard deviations of 3.9 inches (men) and 2.8 inches (women). If Dick and Jane each have normalized heights of $z = 2.3$, what are their respective heights?

Each of them have heights that are 2.3 standard deviations above the means of their respective population. Dick's height is

$$x = \mu + z\sigma = 69.4 + 2.3 \times 3.9 = 78.4$$

or 6 feet 6.4 inches tall. Jane's height is

$$x = \mu + z\sigma = 64.2 + 2.3 \times 2.8 = 70.6$$

or 5 feet 10.6 inches tall.

FAQ 19.  What is z?

# Exercises

Find the $z$ value for each of the following.

1. $x = 33.1$, $\mu = 40$, $\sigma = 11.1$ (ans: $z = -0.62$)
2. $x = 40.73$, $\mu = 52$, $\sigma = 4.6$ (ans: $z = -2.45$)
3. $x = 32.52$, $\mu = 43$, $\sigma = -3.9$ (ans: $z = 2.69$)
4. $x = 52.22$, $\mu = 58$, $\sigma = 8.2$ (ans: $z = -0.7$)
5. $x = 37.92$, $\mu = 41$, $\sigma = 11.1$ (ans: $z = -0.28$)
6. $x = 45.71$, $\mu = 47$, $\sigma = 10.2$ (ans: $z = -0.13$)
7. $x = 47.86$, $\mu = 60$, $\sigma = 11.3$ (ans: $z = -1.07$)
8. $x = 40.17$, $\mu = 45$, $\sigma = 5.7$ (ans: $z = -0.85$)
9. $x = 46.03$, $\mu = 44$, $\sigma = 13.6$ (ans: $z = 0.15$)
10. $x = 62.74$, $\mu = 60$, $\sigma = 4.1$ (ans: $z = 0.67$)
11. $x = 52.8$, $\mu = 43$, $\sigma = 13.4$ (ans: $z = 0.73$)
12. $x = 53.87$, $\mu = 60$, $\sigma = 13.0$ (ans: $z = -0.47$)
13. $x = 38.65$, $\mu = 42$, $\sigma = 15.0$ (ans: $z = -0.22$)
14. $x = 59.66$, $\mu = 59$, $\sigma = 14.8$ (ans: $z = 0.04$)
15. $x = 41.27$, $\mu = 40$, $\sigma = 11.3$ (ans: $z = 0.11$)
16. $x = 48.03$, $\mu = 40$, $\sigma = -2.4$ (ans: $z = -3.35$)
17. $x = 57.01$, $\mu = 48$, $\sigma = -4.9$ (ans: $z = -1.84$)
18. $x = 42.3$, $\mu = 45$, $\sigma = 6.6$ (ans: $z = -0.41$)
19. $x = 55.12$, $\mu = 44$, $\sigma = -4.6$ (ans: $z = -2.42$)

Find $x$ for each of the following $z$ values.

20. $z = -0.24$, $\mu = 55$, $\sigma = 2.8$ (ans: $z = 54.33$)
21. $z = -1.19$, $\mu = 40$, $\sigma = 3.1$ (ans: $z = 36.31$)
22. $z = 0.03$, $\mu = 56$, $\sigma = 5.9$ (ans: $z = 56.18$)
23. $z = 1.14$, $\mu = 47$, $\sigma = -1.9$ (ans: $z = 44.83$)
24. $z = -0.65$, $\mu = 45$, $\sigma = 6.5$ (ans: $z = 40.77$)
25. $z = 0.6$, $\mu = 57$, $\sigma = 14.5$ (ans: $z = 65.7$)
26. $z = 0.7$, $\mu = 47$, $\sigma = 3.8$ (ans: $z = 49.66$)
27. $z = -1.62$, $\mu = 58$, $\sigma = 9.4$ (ans: $z = 42.77$)
28. $z = 2.75$, $\mu = 44$, $\sigma = -4.0$ (ans: $z = 33.0$)
29. $z = 0.35$, $\mu = 59$, $\sigma = 9.8$ (ans: $z = 62.43$)
30. $z = -0.68$, $\mu = 56$, $\sigma = 0.8$ (ans: $z = 55.46$)
31. $z = 2.38$, $\mu = 58$, $\sigma = 6.6$ (ans: $z = 73.71$)
32. $z = -0.72$, $\mu = 53$, $\sigma = -4.1$ (ans: $z = 55.95$)
33. $z = -0.28$, $\mu = 55$, $\sigma = 11.9$ (ans: $z = 51.67$)
34. $z = -1.03$, $\mu = 50$, $\sigma = -1.6$ (ans: $z = 51.65$)
35. $z = -0.91$, $\mu = 55$, $\sigma = 0.6$ (ans: $z = 54.45$)
36. $z = -0.86$, $\mu = 55$, $\sigma = -0.5$ (ans: $z = 55.43$)
37. $z = 1.6$, $\mu = 42$, $\sigma = 8.2$ (ans: $z = 55.12$)
38. $z = 1.53$, $\mu = 44$, $\sigma = 6.6$ (ans: $z = 54.1$)
39. $z = -0.93$, $\mu = 53$, $\sigma = 4.7$ (ans: $z = 48.63$)

# FAQ 20. What is the 68-95-99.7 Rule?

The **68-95-99.7 rule** describes the spread of a **normal distribution**. The rule tells us what percent of the data will have z-scores with $|z| \leqslant 1$ (68%), $|z| \leqslant 2$ (95%), and $|z| \leqslant 3$ (99.7%). This is equivalent to knowing what percentages of the original data fall within one, two, or three standard deviations of the mean.

---

### 68-95-99.7 Rule

Suppose a population is distributed normally with mean $\mu$ and standard deviation $\sigma$. Then:

| Percentage of Data | Raw Data Values | Normalized Data Values |
|---|---|---|
| 68% | are within $\sigma$ of $\mu$ | satisfy $|z| \leqslant 1$ |
| 95% | are within $2\sigma$ of $\mu$ | satisfy $|z| \leqslant 2$ |
| 99.7% | are within $3\sigma$ of $\mu$ | satisfy $|z| \leqslant 3$ |

---

Figure 20.1.: Illustration of the 68-95-99.7 rule. The shaded area indicates the proportion of data within one standard deviation (top), two standard deviations (middle), and three standard deviations (bottom) of the mean.

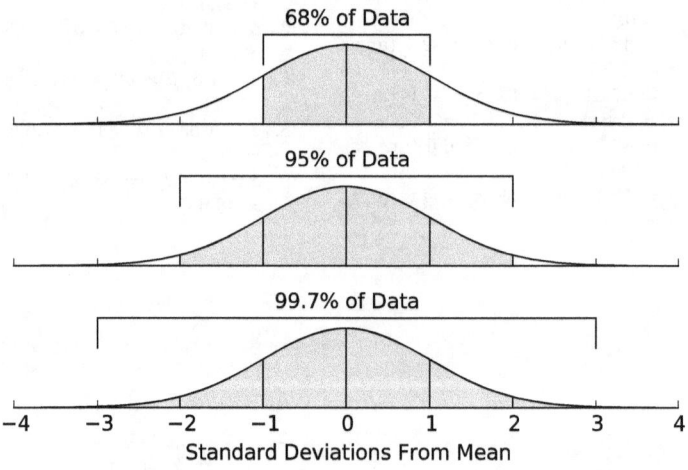

**Example 20.1.** The scores on the entrance level mathematics placement test at a California university are distributed normally with a mean of 50 and standard deviation of 12. Therefore:

- Since $50 - 12 = 38$ and $50 + 12 = 62$, approximately 68% of the scores are between 38 and 62 (within one standard deviation of the mean).
- Since $50 - 2 \times 12 = 26$ and $50 + 2 \times 12 = 74$, approximately 95% of the scores are between 26 and 74 (within two standard deviations of the mean).
- Since $50 - 3 \times 12 = 14$ and $50 + 3 \times 12 = 86$, approximately 99.7% of the scores are between 14 and 86 (within three standard deviations of the mean).

**Example 20.2.** The scores on the entrance level mathematics placement test at a California university are distributed normally with a mean of 50 and standard deviation of 12. What proportion of the scores exceed a value of 62?

Since $62 - 50 = 12$, a score of 62 corresponds to one standard deviation above the mean. By the 68-95-99.7 rule, approximately 68% of the data falls between one standard deviation below the mean and one standard above the mean. Thus the remaining $100 - 68 = 32\%$ of the data is at least one standard deviation from the mean. Half of this, or 16% is at least one standard deviation above the mean, and half is at least one standard deviation below the mean. See figure 20.2.

Figure 20.2.: Illustration of example 20.2. Since 68% of the data lies in the central (shaded) region between 38 and 62, then remaining 32% lies in the two tail areas. Half lies in the upper tail, and half in the lower tail.

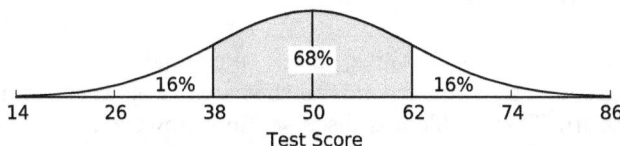

# Exercises

1. The average height of men aged 20-29 in the US is 5'10". The heights are distributed normally with a standard deviation of 2 inches. What proportion of men have height over six feet? over 6'2"? (ans: 16%; 2.5%)

2. The grades on a statistics exam are distributed normally with mean 67 and standard deviation 7. A student must obtain a score of at least 60 to pass the exam. what proportion of students pass the exam? (ans: 84%)

3. Repeat the previous question but assume that the mean of the exam is 74. (ans: 97.5%)

4. The grades on a statistics exam are distributed normally with mean 74 and standard deviation 7. A student who scores at least a 95 will receive a grade of A+. What proportion of students will receive a grade of A+ in this class?

# FAQ 21. How Do I Use a Table of the Normal Distribution?

Figure 21.1.: The shaded area is tabulated in tables of the normal distribution.

Tables of the standard normal distribution, sometimes referred to as z-tables, give the area under the curve of a standard normal distribution **to the left of a value** $z$ (fig. 21.1). This area is sometimes called $P(Z < z)$ because it gives **the proportion of the area that falls to the left of** $z$;

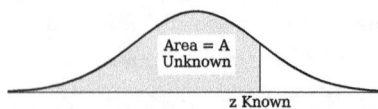

it is also the **probability of an event occurring with normalized value less than** $z$.

A table of the standard normal distribution can be used in two different ways:

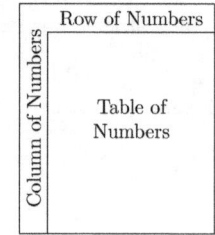

- ▸ Given a $z$ value, use it to find the area (probability) under the curve to the left of that value. This is called a **direct lookup** into the table, and is the subject of this chapter.

- ▸ Given an area (probability), use it to estimate the $z$-value that corresponds to that area. This is called a **reverse lookup.** This problem is discussed in chapter 22.

An example of a typical table is given in appendix A. These tables all have the same general format, as illustrated in the sketch on the right

## Finding the Area to the Left of $z$

The **column of numbers** along the left start with zero and usually increase to 3 or 3.5 in increments of 0.1. This number represents **the integer part** and **first decimal digit** of $z$. Some versions of the table also contain a second page that begins at -3 or -3.4 and increases to 0, along the left column, also in steps of 0.1. (e.g., table A.1).

The **row of numbers** across the top are all two-digit decimals starting with either a decimal point or a zero and a decimal point:

These numbers represent the **second digit to the right of the decimal** of $z$

To find the area corresponding to any value of $z$ using table A.1

1. Round off $z$ to two decimal places.

   **Example 21.1.** If $z = 1.437$, round it to $z = 1.44$. If $z = -2.171$, round it to $z = -2.17$.

2. Write $z$ as the following sum

$$z = \boxed{\text{Truncation to first digit}} + \boxed{\text{Value of second digit}}$$

   **(Continuation of example 21.1.)** $z = 1.44$ becomes

$$z = 1.44 = \boxed{1.4} + \boxed{0.04}$$

   and $z = -2.14$ becomes

$$z = -2.17 = \boxed{-2.1} + \boxed{0.07}$$

3. Look in the table and find the intersection of the row that contains the integer part with the column that contains the hundredths part.

   **(Continuation of example 21.1.)** For $z = 1.44$, we are looking for the intersection of the row containing $\boxed{1.40}$ and the column containing $\boxed{0.04}$. If we look in a table of the normal distribution, it looks something like this (with all the cells filled in with numbers). We follow the arrows indicated here to the intersection of our desired row and column.

| | $\cdots$ | $\cdots$ | 0.03 | 0.04 | 0.05 | $\cdots$ |
|---|---|---|---|---|---|---|
| $\cdots$ | | | | ↓ | | |
| 1.2 | | | | ↓ | | |
| 1.3 | | | | ↓ | | |
| 1.4 | → | → | → | .9251 | | |
| 1.5 | | | | | | |
| 1.6 | | | | | | |
| $\cdots$ | | | | | | |

   The intersection of the row containing $\boxed{1.40}$ and the column containing $\boxed{0.04}$ has the number $\boxed{0.9251}$ in it. Therefore

$$P(z < 1.44) = \text{Area to the left of } 1.44 = 0.9251$$

Similarly, from appendix A, the intersection of the row containing $\boxed{-2.1}$ and the column containing $\boxed{0.07}$ has the number $\boxed{0.0150}$ in it. Therefore

$$P(z < -2.1) = \text{Area to the left of } -2.1 = 0.0150$$

4. If the value you are looking for has $z > 3.49$, then the area is larger than the largest value in the table. This is the the value at the intersection of the largest column (3.4) and largest row (0.09).

**Example 21.2.** Suppose $z = 3.9$. Then

$$P(z < 3.9) = \text{Area to the left of } 3.9 > 0.9998$$

We often express numbers such as this that are "off the right" end of the table as "essentially equal to one."

5. if the value you are looking for has $z < -3.49$, then the area is smaller than the smallest value in the table. This is the value in the intersection of the most negative row (-3.4) and the largest column (0.09).

**Example 21.3.** Suppose $z = -4.2$. Then

$$P(z < -4.2) = \text{Area to the left of } -4.2 < 0.0002$$

We often express numbers such as this that are "off the left" end of the table as "essentially zero."

**Technology Example 21.4 (TI-84).** Find $P(z < 0.8)$ using a TI-84.

1. Select the distributions menu: [2nd] [var].
2. Scroll to **normalcdf(** and press [enter] or press [2].
3. Scroll to **upper:** and type [.] [8].
4. Scroll to **Paste:** and hit [enter]. You will see **normalcdf(-1e99,0.8,0,1)** on your display.
5. Press [enter]. The answer **0.7881446663** will be displayed.

# Finding the Area to the Right of $z$

Finding the area to the right of some value $z$ is a two step process:

1. Find the area $A_L$ to the left of $z$.

---

2. Use the fact that the total area under the curve is 1 to calculate the area to the right as $A = 1 - A_L$.

**Example 21.5.** Find $P(z > 1.44)$. Since we are finding the proportion that is greater than 1.44, this is an area to the right. We first find the area to the left as in example 21.1,

$$A_L = P(z < 1.44) = 0.9251$$

Then the area to the right is

$$P(z > 1.44) = 1 - P(z < 1.44) = 1 - A_L = 1 - .9251 = 0.749$$

**Technology Example 21.6 (TI-84).** Find $P(z > -0.5)$ using a TI-84.

1. Select the distributions menu: [2nd] [var].
2. Scroll to **normalcdf(** and press [enter] or press [2].
3. On older calculators, where "**normalcdf(**" immediately appears on your screen, skip to step 8.
4. Scroll to **lower:** and type [(-)] [.] [5]. (In other words, enter the value $-.5$. Note that there is a single key on the calculator marked [(-)]; do not use the key marked [-] or the parenthesis keys here.)
5. Scroll to **upper:** and type in a large number, such as **1e99**.
6. Scroll to **Paste:** and hit [enter]. You will see **normalcdf(-0.5, 1e99,0,1)** on your display.
7. Press [enter]. The answer **0.6914624678** will be displayed.
8. On older calculators, fill in **normalcdf(-0.5, 1e99, 0, 1)**. (The **1e99** can be any large number; the **0** and **1** are the mean and standard deviation of the standard normal distribution.) Then hit [enter]. The answer **0.6914624678** will be displayed.

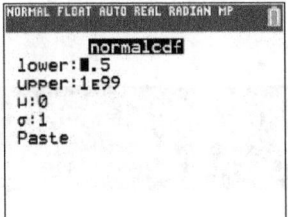 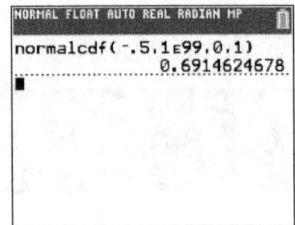

**Technology Example 21.7 (R).** Find $P(z < -0.5)$ using R.

We can do this using the **pnorm** function:

```
> pnorm(-.5)
[1] 0.3085375
```

**Technology Example 21.8 (Spreadsheet).** Find $P(z < -0.5)$ using a spreadsheet.

We can do this in either of the following ways.

1. Using **norm.s.dist(x, cum)** (where **cum=1** is always a flag set equal to 1 to indicate the cumulative density function rather than the density function). To get $P(z < -0.5)$ in this way we would would type
$$=\text{norm.s.dist}(-.05, 1) \enter$$
   into any cell ; or
2. Using **norm.dist(x, mean, sd, cum)** (where **mean** and **sd** are the mean and standard deviation of the normal distribution; and **cum** is always set equal to **1** for a cumulative density function). Since the standard normal distribution has mean 0 and standard deviation 1, to get $P(z < -0.5)$ in this way, type
$$=\text{norm.dist}(-0.5, 0, 1, 1) \enter$$
   into any cell.

# Exercises

Find the following areas by direct table lookup.

1. To the left of $z = .4$ (ans: 0.3446)
2. To the left of $z = -1.3$ (ans: 0.0968)
3. To the right of $z = -.72$ (ans: 0.2358)
4. To the right of $z = 2.85$ (ans: 0.0022)

5. To the right of $z = 3.94$ (ans: < 0.0002)
6. Between $z = -.51$ and $z = .55$ (ans: 0.4038)
7. Between $z = -1.22$ and $z = 0$ (ans: 0.3888)
8. Between $z = 0$ and $z = 2.0$ (ans: 0.4772)
9. Between $z = 0.55$ and $z = 2.35$ (ans: 0.2818)
10. Between $z = -1.97$ and $z = 0.26$ (ans: 0.5782)

# FAQ 22. How Do I Do a Reverse Lookup in a Table of the Normal Distribution?

Suppose we know the area under the curve but don't know the corresponding $z$ value. The problem of finding the $z$ value can be solved by **reverse lookup** into a table of the normal distribution. This is exactly the opposite of the problem that was solved in chapter 21, where we knew

Figure 22.1.: In a reverse lookup, the area is known, and $z$ is unknown. Compare with fig. 21.1.

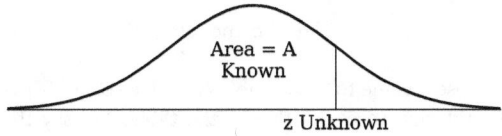

the $z$ value, and used it to look up the the area under the curve to the left of $z$.

To perform a reverse lookup:

1. Determine the "fraction" or convert the "value" you are asked to find into an appropriate "area to the left." Look for key words; for example, "quarter of the data" means 25% or an area to the left of 0.25.

   ▸ Areas to the left can sometimes be flagged by words or expressions like "below," "beneath," "under," "less than," "smaller than," "fewer than," "does not exceed."
   ▸ Areas to the right are are sometimes flagged by expressions like expressions such as "above," "over", "exceeds", "more than", "greater than." The area must be a number between 0 and 1.
   ▸ If the area you are given is an area $R$ to the right, subtract it from one to get an area to the left, $A = 1 - R$
   ▸ Central areas must be converted to areas from the the left. If the central area is $C$, then the area to the left of the left edge is $A = (1 - C)/2$, and the area to the left of the right edge is $A = (1 + C)/2$

2. Locate the area in the table of values for the normal distribution (appendix A). Rather than looking along the edges, we look inside the table, and **find the closest value to the desired area.**

3. Follow the area value you found to the edge of the table to read off the $z$ value.

4. If needed, convert back to raw data ($x$ values). The conversion formula is $x = \mu + z\sigma$.

**Example 22.1.** For what $z$ value does one fifth of the area under the curve lie to the left of $z$?

1. Convert into an area to the left. Since one fifth equal 0.2,

$$\text{Area to the left} = 0.2$$

2. Locate the closest value to 0.2 in table A.1. These are $\boxed{0.1977}$ and $\boxed{0.2005}$, which are right next to each other in the table. We say that these numbers **bracket** 0.2. We observe that $\boxed{0.2005}$ is closer to 0.2 than $\boxed{0.1977}$.
3. Follow $A = .2005$ to the edge of the table to conclude that $z = . - 84$.

**Technology Example 22.2 (TI-84).** Repeat exercise 22.1 using a TI-84.

1. Type $\boxed{\text{2nd}}$ $\boxed{\text{var}}$ and then either scroll to "**3:invNorm(**" or hit $\boxed{3}$.
2. On older calculators the command "**invNorm(**" will be displayed. Skip to step 4.
3. Set **area** to 0.2; leave the other parameters unchanged; scroll to **Paste** and hit $\boxed{\text{enter}}$. You will see **invNorm(.2, 0, 1)** displayed on the screen. Skip to step 5
4. Fill in **invNorm(.2,0,1)**
5. Hit $\boxed{\text{enter}}$ and the answer will be displayed as **−0.8416212335**.

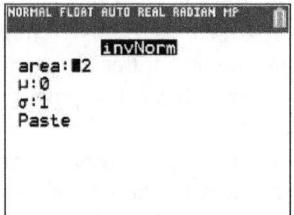 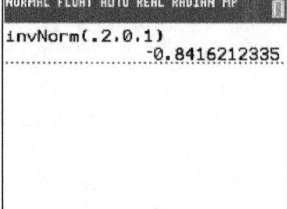

**Technology Example 22.3 (R).** Repeat exercise 22.1 using R.

We can do this with an the **qnorm** function.

```
> qnorm(.2)
[1] -0.8416212
```

**Technology Example 22.4 (Spreadsheet).** Repeat exercise 22.1 using a spreadsheet.

We can do this with an the **norm.inv(p, mean, sd)** function.

1. Type =norm.inv(.2, 0, 1) into any cell. The first number (0.2) is the probability. The second and third numbers (0 and 1) are the mean and standard deviation of a standard normal distribution.
2. Hit enter. The answer will immediately appear in the same cell.

**Example 22.5.** For what $z$ value does three fourths of the area under the curve lie to the right of $z$?

1. We observe the key words **to the right**, and that three fourths of the area must lie to the right. Since $3/4=0.75=R$, the area to the left is $A = 1 - R = 1 - 0.75 = 0.25$.
2. The closest values to 0.25 in A.1 are $\boxed{0.2483}$ and $\boxed{0.2514}$, but $\boxed{0.2514}$ is closest.
3. Follow $A = 0.2514$ to the edges of the table to conclude that $z = -.67$

**Example 22.6.** Find the $z$ value for the 60th percentile.

1. The 60th percentile means that 60% of the data is less than $x$, or when it is normalized, 60% of the data is less than $z$. Thus

$$\text{Area to the left} = 0.6$$

2. $A = 0.6$ is bracket by $\boxed{0.5987}$ and $\boxed{0.6026}$ in table A.1. The first value is closer.
3. Follow 0.5987 to the edges of the table to conclude that $z = 0.25$.

**Example 22.7.** What is the $x$ value corresponding to the third quartile (75th percentile), if $\mu = 47$ and $\sigma = 3$?

1. Convert into an area to the left. The 75th percentie means 75% of the data values fall to the left, so that

$$\text{Area to the left} = 0.75$$

2. $A = 0.75$ is bracketed by $\boxed{0.7517}$ and $\boxed{0.7486}$ table A.1, with $\boxed{0.7486}$ in closer to 0.75.
3. Follow 0.7486 to the edge of the table to conclude that $z = 0.67$
4. Convert to raw data:

$$x = \mu + z\sigma = 47 + (0.67)(3) = 49.01$$

**Technology Example 22.8 (TI-84).** Repeat example 22.7 using a TI-84.

1. Hit 2nd vars and scroll to invNorm or press 3.
2. On older calculators, if "**invNorm(**" appears on screen, skip to step 4.
3. Set the value of **area** to **0.75**; of $\mu$ to 47; and of $\sigma$ to 3. Then scroll to **paste** and press enter. Skip to step 5.

4. Fill in **invNorm(0.75,47,3)**.
5. Press enter. The answer **49.02346925** will appear on the screen. The difference with answer in example 22.7 is due to roundoff error.

**Technology Example 22.9 (R).** Repeat example 22.7 using R. We can do this using the function **qnorm(p, mean=value, sd=value))**.

```
> qnorm(.75, mean=47, sd=3)
[1] 49.02347
```

**Example 22.10.** What are the $x$ values corresponding to the central fifty percent of the data when $\mu = 10.5$ and $\sigma = 2$?

1. The central 50 percent of the data corresponds to the data that falls between the 25th and 75th percentile This means that we need to find two values. There are two "areas to the left." They are

$$\text{Area to the left} = 0.25$$

and

$$\text{Area to the left} = 0.75$$

2. $A = 0.25$ is bracketed by $\boxed{0.2483}$ and $\boxed{0.2514}$, with $\boxed{0.2514}$ closer to 0.25.
3. Following 0.2514 to the edge of the table gives $z = -0.67$
4. $A = 0.75$ is bracketed by $\boxed{0.7517}$ and $\boxed{0.7486}$, with $\boxed{0.7486}$ closer to 0.75.
5. Following $A = 0.75$ to the edge of the table gives $z = 0.67$
6. We convert the standardized values $z = -0.67$ and $z = 0.67$ back to raw data values.

$$x = \mu - \sigma z = 10.5 - 2(0.67) = 9.16 \qquad \text{(25th percentile)}$$
$$x = \mu + \sigma z = 10.5 + 2(0.67) = 11.84 \qquad \text{(75th percentile)}$$

The central 50% lies between 9.16 and 11.84.

# Exercises

Find the z value corresponding to the following area by reverse table lookup.

1. Area $= 0.5000$ (ans: $z = 0$)
2. Area $= 0.7500$ (ans: $z = 0.67$)
3. Area $= 0.6000$ (ans: $z = 0.25$)
4. The $90^{th}$ percentile (ans: $z = 1.28$)
5. The $45^{th}$ percentile (ans: $z = -0.13$)
6. The $10^{th}$ percentile (ans: $z =$

−1.28)

7. If a statistics exam has mean 70 and standard deviation 10, what grade is necessary to be in the top 10 percent (round up to the nearest integer)? (ans: 83)

8. Suppose a student needs to be at least in the $45^{th}$ percentile to pass a statistics exam. If the exam has a mean 70 and standard deviation 10, what score is necessary to pass? (Round to the nearest integer.) (ans: 69)

## FAQ 23. How Do I Find a Probability for a Normal Distribution?

For a normal distribution, the following numbers are the same:

▸ The area under the curve of the distribution between two points $a$ and $b$.
▸ The proportion of individuals in the population whose values are between two numbers $a$ and $b$.
▸ The probability that an individual selected at random from the the population will have a value between the numbers $a$ and $b$.

We compute these areas, or proportions, or probabilities, by standardization (conversion to $z$ value). As discussed below (in chapter 26),

$$\text{Probability}(a < x < b) = \text{Probability}\left(\frac{a - \mu}{\sigma} < z < \frac{b - \mu}{\sigma}\right)$$
$$= \text{Area}\left(\frac{a - \mu}{\sigma} < z < \frac{b - \mu}{\sigma}\right)$$
$$= \text{Area}\left(z < \frac{b - \mu}{\sigma}\right) - \text{Area}\left(z < \frac{a - \mu}{\sigma}\right)$$

The following three numbers are also equivalent:

▸ The area under the curve of the distribution to the left of any number $a$.
▸ The proportion of individuals in the population whose value is smaller than $a$.
▸ The probability that an individual selected at random from the the population will have a value less than $a$.

Figure 23.1.: Area under a distribution is equivalent to probability.

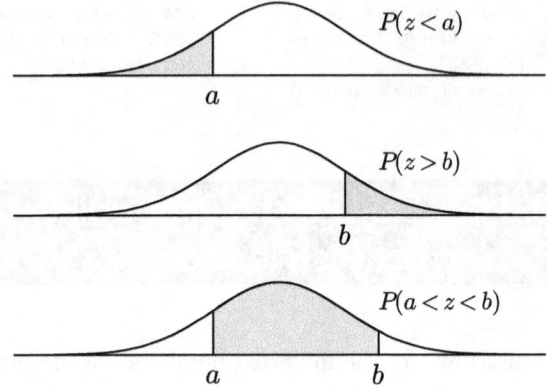

As will be discussed in more detail in see chapter 24,

$$\text{Probability}(x < b) = \text{Probability}\left(z < \frac{b - \mu}{\sigma}\right) = \text{Area}\left(z < \frac{b - \mu}{\sigma}\right)$$

Finally, the following three numbers are equivalent:

- The area under the curve of the distribution to the right of any number $b$.
- The proportion of individuals in the population whose value is larger than $b$.
- The probability that an individual selected at random from the the population will have a value greater than $b$.

This will be discussed in 25,

$$\text{Probability}(x > a) = \text{Probability}\left(x > \frac{a - \mu}{\sigma}\right)$$
$$= \text{Area}\left(z > \frac{a - \mu}{\sigma}\right)$$
$$= 1 - \text{Area}\left(z < \frac{a - \mu}{\sigma}\right)$$

The formulas given above are summarized in appendix C and are explored in more detail in chapters 24, 25, and 26.

# FAQ 24. How Do I Find the Probability that a single individual x is smaller than something?

Suppose we select a single individual at random from a popuation that has a normal distribution with mean $\mu$ and standard deviation $\sigma$.

The probability $P(x < A)$ that $x < A$ is equal to the area under the density curve **to the left** of $A$. Since the item is drawn from a population that is normally distributed with mean $\mu$ and standard deviation $\sigma$, the raw data value $A$ can be normalized using the z value (see chapter 19)

$$a = \frac{A - \mu}{\sigma}$$

As discussed in chapter 23 the probability that $x < A$ is the same as the probability that $z < a$, where $z$ is drawn from a standard normal distribution with mean 0 and standard deviation 1 (fig. 24.1).

$$P(x < A) = P(z < a) = P\left(z < \frac{A - \mu}{\sigma}\right)$$

We will take advantage of this equality to compute the probability. We can do this because the areas under standardized normal curves are tabulated.

Figure 24.1.: Areas under raw distribution curves give proportions; the area to the left of $x = A$ and under the curve, divided by the total area under the curve, gives the proportion of data in the population that satisfies $x < A$. The area under the normalized distribution curve (right) sums to one, so that proportional areas give probabilities. The curve on the right is centered at zero, rather than at $\mu$.

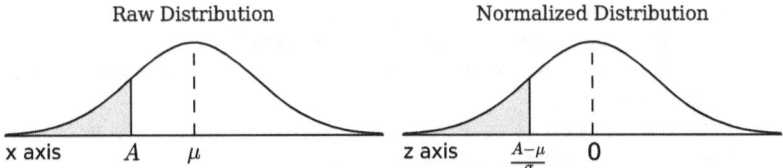

To find the Probability to the Left (fig. 24.2),

Figure 24.2.: The probability is computed first by normalizing the raw value, and then by finding the appropriate area under the normal curve.

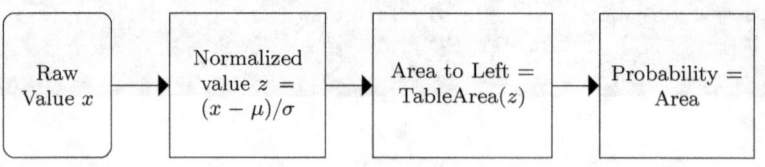

1. Given $x = A$, find the corresponding z-value; call this z-value $z = a$:

$$a = \frac{A - \mu}{\sigma}$$

**Example 24.1.** Suppose that the national distribution of Mathematics SAT scores this year has a mean of 470 and a standard deviation of 140. We are interested in finding the probability of scoring under 400, $P(x < 400)$. Then the normalized value corresponding to $A = 400$ is

$$a = \frac{400 - 470}{140} = -0.5$$

2. Since $P(x < A) = P(z < a)$, find the area under the curve of a standard normal distribution to the left of $z = a$ either using software or a table (such as table A.1).

**(Continuation of example 24.1.)** The area under the curve to the left of $z = a = -0.5$ is $p = 0.3085$. This is also the probability that $z < a = -0.5$ or equivalently, the probability that $x < A = 400$. Thus

$$P(x < 400) = 0.3085 = 30.85\%$$

**Technology Example 24.2 (TI-84).** Repeat example 24.1 using the TI-84. Use **normalcdf(lower,upper,mean,sd)** function (see, e.g., example 21.4).

1. Select the distributions menu: [2nd] [vars].
2. Scroll to **normalcdf(** and press [enter] or press [2].
3. Leave **lower** set to a large negative number. This means you are finding an area to the left.
4. Set **upper** to **400**. This is the upper bounds.
5. Set $\mu$ to **470**. This sets the mean value.
6. Set $\sigma$ to **140**. This sets the standard deviation.
7. Scroll to **Paste** and hit [enter]. When you see **normalcdf(-1e99,400,470,140)** press [enter]. The answer is displayed on the screen:

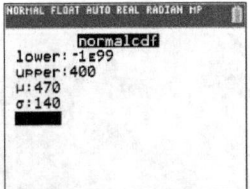
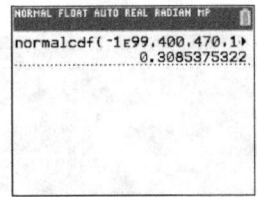

**Technology Example 24.3 (R).** Repeat example 24.1 using R.

Do this using the **pnorm(x, mean, sd)** function.

```
> pnorm(400,mean=470,sd=140)
[1] 0.3085375
```

**Technology Example 24.4 (Spreadsheet).** Repeat example 24.1 using a spreadsheet.

Do this with the **norm.dist(x, mean, sd, cum)** function (see example 21.8).

# Exercises

Find the specified probability for the given normal distributions.

1. $P(x < 23)$ when $x$ is $N(25, 2)$.
   (ans: $p = 0.1587$, software)
2. $P(x < 24)$ when $x$ is $N(25, 2)$.
   (ans: $p = 0.3085$, software)
3. $P(x < 24)$ when $x$ is $N(25, 5)$.
   (ans: $p = 0.4207$, software)
4. $P(x < 20.7)$ when $x$ is $N(23.6, 4.2)$. (ans: $p = 0.2449$, software)
5. $P(x < 250)$ when $x$ is $N(510, 95)$.
   (ans: $p = 0.0031$, software)
6. $P(x < 400)$ when $x$ is $N(496, 75)$.
   (ans: $p = 0.1003$, software)

In the following exercises, assume that the variables are are distributed normally.

7. Suppose a fleet of cars has a mean highway gas mileage of 27 miles per gallon, with standard deviation 2.5 miles per gallon. What is the probability that a randomly selected car has highway mileage less than 25 mpg? less than 20 mpg? less than 30 mpg? (ans: 0.2119, 0.0026, 0.8849 )

8. The SAT exam has a mean of 500 and standard deviation of 100 What is the probability of scoring less than a 250 on the SAT Math test? below a 400? below a 700? (ans: 0.0062, 0.1587, 0.9772 )

9. A statistics teacher tells the class that students whose final exam grades are 60 and below will fail the class. If the class mean in 67 and the standard deviation 17, what is the probability that a student will fail the class (ans: 0.3403)

---

# FAQ 25. How Do I Find the Probability that a Single Individual x is Bigger than Something?

Suppose we select a single individual at random from a popuation that has a normal distribution with mean $\mu$ and standard deviation $\sigma$.

The probability that $x > B$ is equal to the area under the density curve **to the right** of $B$. If the data is assumed to drawn from a population that is normally distributed with mean $\mu$ and standard deviation $\sigma$, the raw data value $A$ can be normalized to

$$b = \frac{B - \mu}{\sigma}$$

As discussed in chapter 23, the probability that $x > B$ is the same as

Figure 25.1.: Areas under raw distribution curves give proportions; the area to the right of $x = B$ and under the curve, divided by the total area under the curve, gives the proportion of data in the population that satisfies $x > B$. The area under the normalized distribution curve (right) sums to one, so that proportional areas give probabilities. The curve on the right is centered at zero, rather than at $\mu$. The area to right of $(B-\mu)/\sigma$ in the curve on the right is the probability that $x > B$.

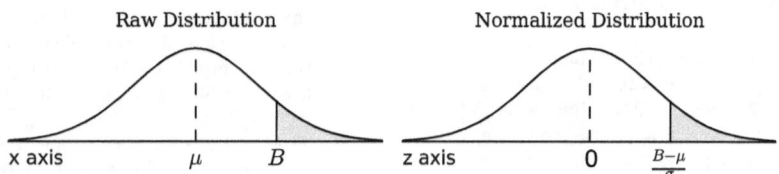

| Raw Distribution | Normalized Distribution |
|---|---|
| x axis    $\mu$   $B$ | z axis    $0$   $\frac{B-\mu}{\sigma}$ |

the probability that $z > b$, where $z$ is drawn from a standard normal distribution with mean 0 and standard deviation 1 (fig. 25.1)

$$P(x > B) = P(z > b) = P\left(z > \frac{B - \mu}{\sigma}\right)$$

We will take advantage of this equality to compute the probability. We can do this because the areas under standardized normal curves are tabulated. Because areas under the curve are only tabulate to the left of a value, this is one additional step here than in chapter 24.

Figure 25.2.: Summary of procedural flow for calculating the probability that an individual selected at random is larger than a specific value.

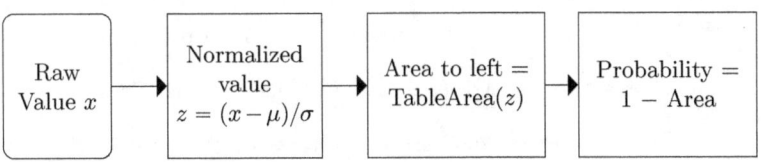

To find $P(x > A)$, the area to the right of $A$, is:

1. Calculate the normalized value $b$ corresponding to the raw data value $B$:

$$b = \frac{B - \mu}{\sigma}$$

**Example 25.1.** Suppose that the national distribution of Mathematics SAT scores this year has a mean value of 470 and a standard deviation of 140 this year. We are interested in finding the probability of scoring over 700, $P(> 700)$. This means that $\mu = 470$, $\sigma = 140$, and $B = 700$. Then the normalized value $b$ is

$$b = \frac{700 - 470}{140} = 1.64$$

2. Find the area under the curve of a standard normal distribution to the right of $z = b$ either using software or a table lookup (such as table A.1; table lookup is described in chapter 21). This works because $P(x > B) = P(z > b)$.

**(Continuation of example 25.1.)** We find the area to the right of $z = b = 1.64$ in three different ways.

a) Use a table, and look up the area under the curve to the left of $z = -n$, which we call $\mathrm{Area}(z < -b)$. This gives us the area to the left of $-b$. Because of the symmetry properties of the standard normal curve, the area under the curve to the right of $z = b$ is the same as the area under the curve to the left of $z = -b$. From table A.1, we find that this is 0.505. Therefore,

$$P(x > B) = \mathrm{Area}(z > b) = \mathrm{Area}(z < -b) = 0.0505$$

Thus the area to the left of $z = -1.64$ is 0.0505.

b) Use a table (such as table A.1), and look up the area under the curve to the left of $z = b$. We will call this Area($z < b$).

You can't look up the area to the right of $z = b$ because this type of table only gives the area to the left of $b$. Unfortunately, what you want is the area to the right of $z = b$. Luckily, you know that the total area under the curve is 1. Then

$$P(x > B) = \text{Area}(z > b) = 1 - \text{Area}(z < b)$$

Table A.1 gives Area($z > b$) = .9495 for $z = 1.64$. Thus

$$P(x > 700) = P(z > 1.64) = 1 - \text{Area}(z < 1.64)$$
$$= 1 - .9495 = 0.0505$$

**Technology Example 25.2 (TI-84).** Repeat example 25.1 using a TI-84.

Compare with example 24.2. Use the **normalcdf(lower,upper,mean,sd)** function (see, e.g., example 21.4):

1. Select the distributions menu: [2nd] [vars].
2. Scroll to **normalcdf(** and press [enter] or press [2].
3. Set **lower** to **700**. This is the lower bounds.
4. Set **upper** to a large positive number such as **1e99** or **9999999**. This means you are finding the entire area to the right.
5. Set $\mu$ to **470**. This sets the mean value.
6. Set $\sigma$ to **140**. This sets the standard deviation.
7. Scroll to **Paste** and hit [enter]. When you see **normalcdf(700,1e,99,470,140)**, press [enter]. The answer is displayed on the screen:

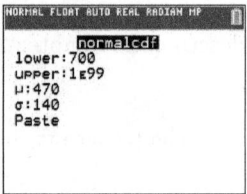
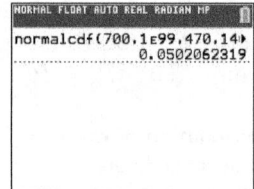

**Technology Example 25.3 (R).** Repeat example 25.1 using R.

Do this with the **pnorm(x, mean, sd, lower.tail=FALSE)** function. Setting **lower.tail=FALSE** means find the area to the right instead of the area to the left. Compare with example 24.3.

```
> pnorm(700,mean=470,sd=140,lower.tail=FALSE)
[1] 0.05020625
```

**Technology Example 25.4 (Spreadsheet).** Repeat example 25.1 using a spreadsheet.

The **norm.dist(x, mean, sd, cum)** function (see example 24.4) gives the area to the left of **x**; to get the area to the right, we subtract from one.

| *fx* | =1-NORM.DIST(700,470,140,1) | | |
|---|---|---|---|
| | C | D | E |
| | 0.05020625 | | |

# Exercises

variables are are distributed normally.

Find the specified probability for the given normal distributions.

1. $P(x > 23)$ when $x$ is $N(25, 2)$.
   (ans: $p = 0.8413$, software)
2. $P(x > 26)$ when $x$ is $N(25, 2)$.
   (ans: $p = 0.3085$, software)
3. $P(x > 30)$ when $x$ is $N(25, 5)$.
   (ans: $p = 0.1587$, software)
4. $P(x > 90)$ when $x$ is $N(77, 12.)$.
   (ans: $p = 0.1393$, software)
5. $P(x > 65)$ when $x$ is $N(73, 15.)$.
   (ans: $p = 0.7031$, software)
6. $P(x > 250)$ when $x$ is $N(510, 95)$.
   (ans: $p = 0.9969$, software)
7. $P(x > 700)$ when $x$ is $N(475, 80)$. (ans: $p = 0.0025$, software)
8. $P(x > 400)$ when $x$ is $N(496, 75)$.
   (ans: $p = 0.8997$, software)

In the following exercises, assume that the

9. Suppose a fleet of cars has a mean highway gas mileage of 27 miles per gallon, with standard deviation 2.5 miles per gallon. What is the probability that a randomly selected car has highway mileage more than 25 mpg? more than 20 mpg? more than 30 mpg? (ans: .7881, .9974, .1151, software)

10. The SAT exam has a mean of 500 and standard deviation of 100 What is the probability of scoring more than a 250 on the SAT Math test? more than 400? above a 700? (ans: .9938, .8413, .0228, software)

11. A statistics teacher tells the class that students whose final exam grades are 90 and above will get an A in the class. If the class mean in 67 and the standard deviation 17, what is the probability that a student will get an A in the class? (ans: 0.0880, software)

# FAQ 26. How Do I Find the Probability that a Single Individual x is Between Two Values?

Suppose we select a single individual at random from a popuation that has a normal distribution with mean $\mu$ and standard deviation $\sigma$.

The probability that $A < x < B$ is equal to the area under the density curve **between** $A$ and $B$. If the data is assumed to drawn from a population that is normally distributed with mean $\mu$ and standard deviation $\sigma$, the raw data values $A$ and $B$ can be normalized to

$$a = \frac{A - \mu}{\sigma} \text{ and } b = \frac{B - \mu}{\sigma}$$

As discussed in chapter 23, the probability that $A < x < B$ is the same

Figure 26.1.: Areas under raw distribution curves give proportions; the area between $x = A$ and $x = B$ and under the curve, divided by the total area under the curve, gives the proportion of data in the population that satisfies $A < x < B$. The area under the normalized distribution curve (right) sums to one, so that proportional areas give probabilities. The curve on the right is centered at zero, rather than at $\mu$. The area to between of $(A - \mu)/\sigma$ and $(B - \mu)/\sigma$ under the curve on the right is the probability that $A < x < B$.

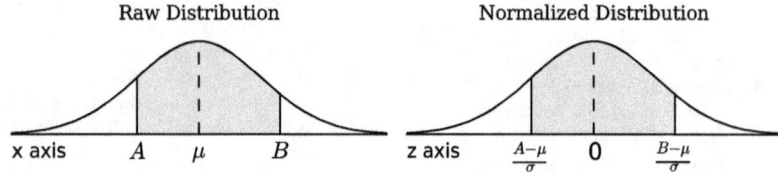

as the probability that $a < z < b$, where $z$ is drawn from a standard normal distribution with mean 0 and standard deviation 1 (fig.26.1).

$$P(A < x < B) = P(a < z < b) = P\left(\frac{A - \mu}{\sigma} < z < \frac{B - \mu}{\sigma}\right)$$

The probability is illustrated by the filled areas in figure 26.1.

**Example 26.1.** Suppose that the Verbal SAT Test scores are normally distributed with mean 483 and standard deviation 72. What is the probability that a randomly selected student has a score that falls between 600 and 700? (to be continued).

To find $P(A < x < B)$, the probability that $x$ falls between the numbers $A$ and $B$, when $x$ is randomly drawn from a normal population, is calculated as follows.

1. Verify that $A < B$. If not, one of the following is possible:

---

FAQ 26.  Probability$(A < x < B)$ when $x$ is Normal

Figure 26.2.: Summary of procedural flow for calculating the probability that an individual selected at random is between two specific values $A$ and $B$. Here $A_L$ is the area under the curve to the left of $A$, and $A_R$ is the area under the curve to the left of $B$.

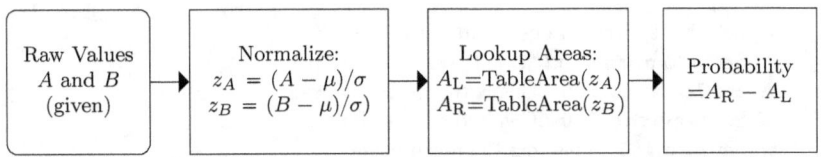

a) You have mislabeled your variables, in which case exchange the values of $A$ and $B$
b) You have labeled the variables correctly, in which case you are asking a question for which the answer domain is empty.

(Continuation of example 26.1.) In this example, $A = 600 < 700 = B$, so we may proceed.

2. Calculate the normalized values of both $A$ and $B$.

$$a = \frac{A - \mu}{\sigma}, \qquad b = \frac{B - \mu}{\sigma}$$

(Continuation of example 26.1.)

$$a = \frac{600 - 483}{72} = 1.625 \text{ and } b = \frac{700 - 483}{72} = 3.014$$

3. Find Area($z < a$) by table lookup (see chapter 21).

(Continuation of example 26.1.)
Using table A.1, TableArea($z < 1.625$) = 0.9479.

4. Find Area($z < b$) by table lookup (chapter 21).

(Continuation of example 26.1.)
Using table A.1, TableArea($z < 3.014$) = 0.9987.

5. $P(A < x < B) = \text{Area}(z < b) - \text{Area}(z < a)$

(Continuation of example 26.1.)

$$P(600 < x < 700) = 0.9987 - .9479 = 0.0508$$

There is about a 5% chance of falling in the range 600 to 700.

**Technology Example 26.2 (TI-84).** Repeat example 26.1 using a TI-84. Use the **normalcdf(lower,upper,mean,sd)** function; in this case, we set both the **lower** and **upper** values.

1. Select the distributions menu: [2nd] [vars].
2. Scroll to **normalcdf(** and press [enter] or press [2].
3. Set **lower** to **600**. This is the lower bounds.
4. Set **upper** to **700**. This is the upper bounds.
5. Set $\mu$ to **483**. This sets the mean value.
6. Set $\sigma$ to **72**. This sets the standard deviation.
7. Scroll to **Paste** and hit [enter]. When you see **normalcdf(600, 700, 483, 72)**, press [enter]. The answer is displayed on the screen:

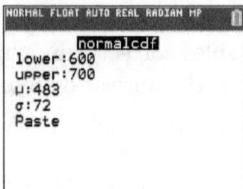 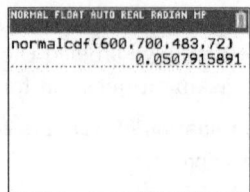

**Technology Example 26.3 (R).** Repeat example 26.1 using R.

We can do this with **pnorm(x, mean, sd)** by precisely emulating the subtraction (larger area) minus (smaller area) that we performed in example 26.1.

```
> pnorm(700,mean=483,sd=72)-pnorm(600,mean=483,sd=72)
[1] 0.05079167
```

**Technology Example 26.4 (Spreadsheet).** Repeat example 26.1 using a spreadsheet.

We can use the function **norm.dist(x, mean, sd, cum)** function to precisely emulate the subtraction of areas we did in example 26.1.

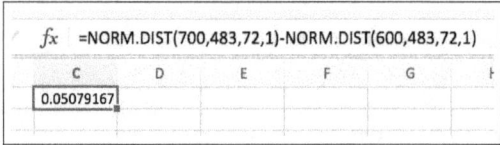

# Exercises

Find the specified probabilities.

1. $P(1 < x < 2)$ when $x$ is $N(5,3)$. (ans: 6.74%)
2. $P(60 < x < 70)$ when $x$ is $N(75,10)$. (ans: 24.17%)
3. $P(60 < x < 70)$ when $x$ is $N(80,10)$. (ans: 13.59%)
4. $P(-5 < x < 2.7)$ when $x$ is $N(1.3,4)$. (ans: 53.92%)
5. $P(550 < x < 700)$ when $x$ is $N(510,95)$. (ans: 31.41%)

In the following exercises, assume that the variables are are distributed normally.

6. Suppose a fleet of cars has a mean highway gas mileage of 27 miles per gallon, with standard deviation 2.5 miles per gallon. What is the probability that a randomly selected car has highway mileage between 25 and 30 mpg? (ans: 67.31%)

7. The SAT exam has a mean of 500 and standard deviation of 100 What is the probability of scoring between 500 and 600? between 500 and 700? between 450 and 650? (ans: 34.13%, 47.72%, 62.47% )

8. A statistics teacher tells the class that students whose final exam scores are 70 and below 80 will get a C, and that scores above 80 and below 90 will result in B. If the class mean in 67 and the standard deviation is 17, what is the probability that a student will get a B? a C? (ans:13.42%, 20.77%)

# FAQ 27. When Do I Subtract from One?

Tables of the normal distribution, such as table A.1, give the area under a curve of a normal distribution to the left a specific value of $z = z_0$. We can call this TableArea($z$)

Figure 27.1.: The area to the left of $z = z_0$ represents the proportion of observations in a normally distributed population that have $z = (x - \mu)/\sigma$ smaller than $z_0$. It also represents the probability that a randomly selected individual from the population with have a normalized value $z = (x - \mu)/\sigma < z_0$. This area is read directly from a table of the normal distribution such as table A.1.

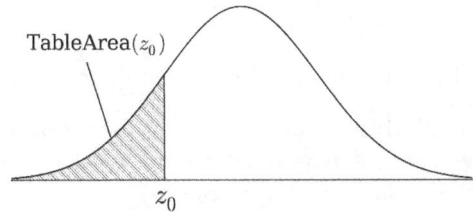

The proportion of observations with $z$ less that $z_0$ is equal to TableArea($z_0$)

(see fig. 27.1):

$$\text{Proportion}(z < z_0) = \text{TableArea}(z_0)$$

The probability that a **single individual** that is randomly selected from the population will have value $x < x_0$, is given by the table vaue of $z = (x_0 - \mu)/\sigma$:

$$\text{Probability}(x < x_0) = \text{TableArea}\left(\frac{x_0 - \mu}{\sigma}\right)$$

If a **random sample of** $n$ **individuals** is selected from the population, the mean of that sample is $\bar{x}$. The probability that $\bar{x}$ is smaller than some value $a$ is given by the table value of $z = (a - \mu)/(\sigma/\sqrt{n})$.

$$\text{Probability}(\bar{x} < a) = \text{TableArea}\left(\frac{a - \mu}{\sigma/\sqrt{n}}\right)$$

The proportion of observations with $z$ larger than some number $z_0$ is

Figure 27.2.: The area to the right of $z = z_0$ represents the proportion of observations in a normally distributed population that have $z = (x - \mu)/\sigma$ larger than $z_0$. This area cannot be read directly from a table of the normal distribution such as table A.1, which gives only the area to the left $z_0$.

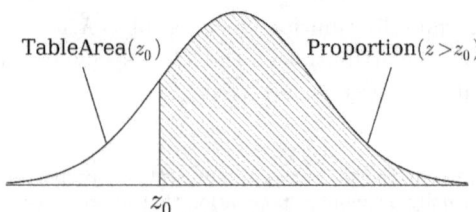

different from the TableValue($z_0$) (see fig. 27.2). In fact, since the total area under the curve is equal to one, the two numbers must add up to one. Hence the area to the right is equal to one minus the area to the left, or one minus the table value.

$$\text{Proportion}(z > z_0) = 1 - \text{TableArea}(z_0)$$

Similarly, **the probability that a single individual that is randomly selected** from the population will have value $x > x_0$, is given by the one minus the table vaue of $z = (x_0 - \mu)/\sigma$:

$$\text{Probability}(x > x_0) = 1 - \text{TableArea}\left(\frac{x_0 - \mu}{\sigma}\right)$$

FAQ 27.   When Do I Subtract from One?

If a **random sample of** $n$ **individuals is selected from the population,** the mean of that sample is $\bar{x}$. The probability that $\bar{x}$ is larger than some value $a$ is given by one minus the table value of $z = (a - \mu)/(\sigma/\sqrt{n})$

$$\text{Probability}(\bar{x} > a) = 1 - \text{TableArea}\left(\frac{a - \mu}{\sigma/\sqrt{n}}\right)$$

To determine whether or not to subtract, you heed to determine whether or not you are finding a probability or proportion of something that is smaller or larger than some specific value. Look for synonyms in the word problems.

In general, synonyms of *larger* give a clue that you will need to subtract from one: bigger, greater, longer, exceeds, bulkier, heavier, more, higher, taller, larger, better, greater, above, over.

In general, synonyms of *smaller* mean that you probably do not need to subtract from one: less, lower, shorter, under, fewer, lesser, lighter, lower, does not exceed, under, below, beneath.

Be careful to read the entire problem. Words like bigger or smaller, and their synonyms, must be read in the context of the problem; the presence of the word alone is not sufficient to make a conclusion as to whether or not you should subtract from one.

## FAQ 28. What is the Central Limit Theorem?

Suppose we draw random samples of size $n$ from a large population of test scores, as schematized in figure 28.1. We then compute the mean of each sample. We are interested in how these means are distributed. This distribution will change with sample size.

To understand this, consider the left hand hand side of fig. 28.1, where we illustrate the collection of random samples of Math SAT test scores. Each sample consists of twelve SAT scores. We then compute the mean of each sample. The first sample has a mean of 516, the second sample has a mean of 502, the third sample has a mean of 502, and so forth. The numbers 516, 502, 502, ... are called the **sample means**. We can represent the sample means symbolically as $\bar{x}_1, \bar{x}_2, \bar{x}_3, \ldots$ If we were to make a histogram of these means it would represent the **distribution of the sample sample means**. We can then ask the following question: what are the center and spread of the disribution of the sample means?

The answer to this question is given by the **Central Limit Theorem**. The answer is only approximate, and only becomes really accurate as $n$, the size of the sample becomes large. However, it is reasonably accurate even for relatively small samples. The Central Limit Theorem is stated in the following box.

---

### Central Limit Theorem

If for some population with mean $\mu$ and standard deviation $\sigma$,

1. A large number of simple random samples of size $n$ are drawn from that population;
2. The **sample means** $\bar{x}_1, \bar{x}_1, \bar{x}_2, \bar{x}_3, \ldots$ of each sample are calculated

then if $n$ is large enough, the **distribution of the sample means** will

1. be approximately normal;
2. have mean $\mu_{\bar{x}} = \mu$
3. have standard deviation $s_{\bar{x}} = \sigma/\sqrt{n}$.

---

Fig. 28.2 illustrates this variation by comparing two different collections of test score samplings. In each case, random samples of size $n$ are selected from national distributions of standardized test scores. These scores are distributed normally with a mean of $\mu = 500$ points and a standard deviation of $\sigma = 80$ points. This is similar to college board exams like the SAT, although the SAT has a larger standard deviation and scores are clipped with a minimum of 200 and maximum of 800.

The scale at the bottom of fig. 28.2 shows the ranges of the scores. Each sample is displayed along a horizontal line, with each test scores in the sample plotted as a small black dots. The mean of each sample is also indicated with a triangle. In the 50 samples on the left, the sample size was ten (i.e., $n = 10$). In the 50 samples on the right, the sample size was 40 (i.e., $n = 40$). Observe how the means are all centrally located. Furthermore, the means are more narrowly distributed on the right than on the left. In fact, the spread of the distribution of the triangles on the right is approximately half the spread of the distribution of the of triangles on the left. Summary statistics are given in the table 28.1.

The histograms of the means are shown in fig. 28.3. The top two frames

---

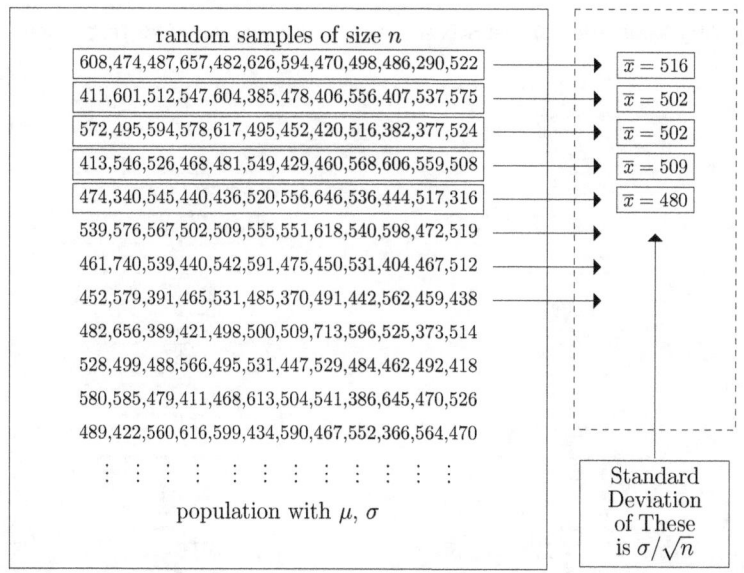

Figure 28.1.: Samples of size $n$ are draw from some large population of test scores with mean $\mu$ and standard deviation $\sigma$. The central limit theorem tells us that for sufficiently large $n$, the distribution of the sample means will have mean $\mu$ and standard deviation $\sigma/\sqrt{n}$.

give the normalized histograms of the 100 sample means of size 10 and size 40 respectively. The bottom frame gives the histogram of 10,000 random test scores from the original population. Each histogram is normalized to give the same total area so that they can be easily compared. The smooth curve in the bottom frame is $N(\mu, \sigma) = N(500, 80)$. According to the central limit theorem, for sufficiently large sample size, this distributions in each of the top two frames should be $N(\mu, \sigma/\sqrt{n})$. These are the curves that are plotted in these two frames.

When you normalize a statistic to its corresponding $z$ value, you must first determine what that statistic represents, because different statistics will have different normalizations. This is because the normalization to $z$ value always measures the distance from mean in units of the standard deviation, and the standard deviation is different for different statistics. If the original statistic is

- **an individual,** such as $x$, then its standard deviation is $\sigma$
- **a sample mean,** such as $\overline{x}$, then its standard deviation is $\sigma/\sqrt{n}$

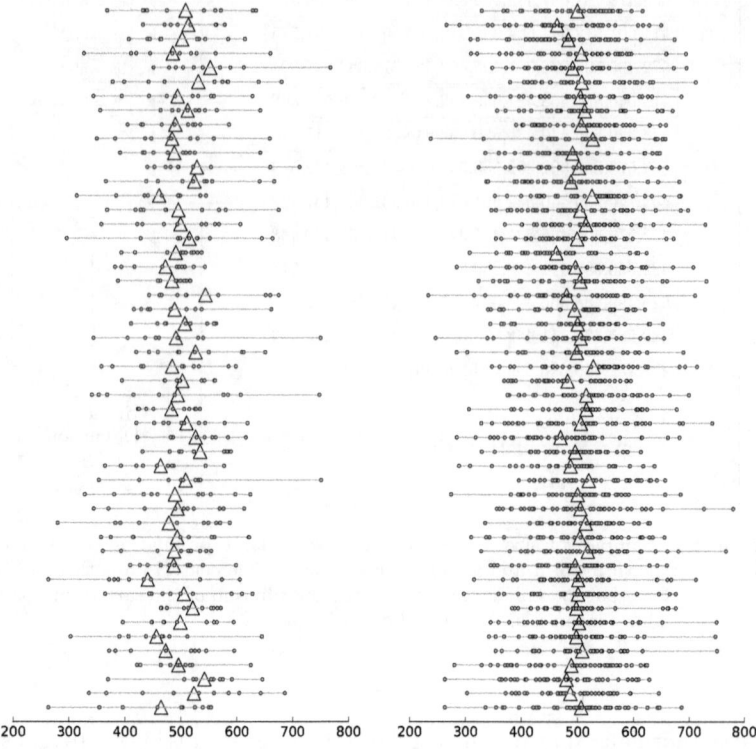

Figure 28.2.: Results of two different collections of test score samples. See the text for details.

Therefore

▸ The *z*-normalization of an individual is called the **z-score**:

$$z = \frac{x - \mu}{\sigma}$$

▸ The *z*-normalization of a sample mean is called the **one sample z-statistic**:

$$z = \frac{\bar{x} - \mu}{\sigma/\sqrt{n}}$$

The distribution of the sample means is corresponding narrower than the population distribution (fig. 28.4). As the sample size gets larger, the

FAQ 28. What is the Central Limit Theorem?

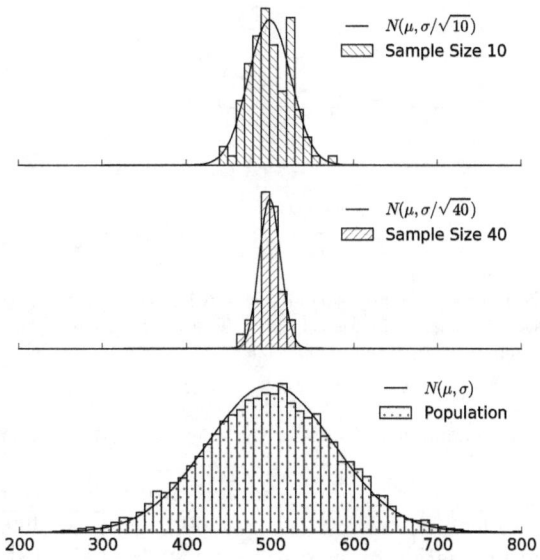

Figure 28.3.: Histograms of the means of the samples shown in fig. 28.2. Top: sample size 10; middle: sample size 40; bottom: full population. The curves give the corresponding normal distributions. See text for details.

spread of the distribution of sample means gets smaller. This is why the triangles on the right hand side of fig. 28.2 are less spread out then the triangles on the left hand side of the same figure. The spread of the distribution is $\sigma/\sqrt{n}$. In other words, approximately 68% of the sample means fall within $\sigma/\sqrt{n}$ of $\mu$, approximately 95% of the sample means fall within $2\sigma/\sqrt{n}$, and approximately 99.7% of the sample means fall within $3\sigma/\sqrt{n}$ of $\mu$.

**Example 28.1.** Suppose that the height of American males in their twenties is normally distributed with mean $\mu = 70$ inches and standard deviation 2.1 inches. Find (a) The probability that a single individual, chosen at random, is over 6'1" tall; and (b) The probability that the average height of a sample of 25 men is over 6'1".

a) We first convert 6'1" to 73". The corresponding $z$-value is

$$z = \frac{x - \mu}{\sigma} = \frac{73 - 70}{2.1} = 1.43$$

Since we are looking for the probability that the height is greater than 6'1".

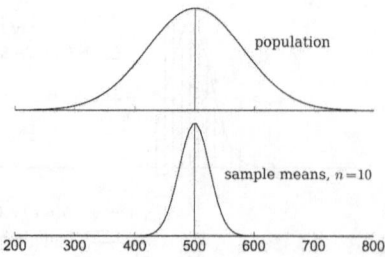

population

sample means, $n=10$

| 200 | 300 | 400 | 500 | 600 | 700 | 800 |

Figure 28.4.: Comparison of population distribution (top) with distribution of the sample means ($\bar{x}$) for $n = 10$. The distribution of the sample means is narrower by a factor of $1/\sqrt{n}$.

Table 28.1.: Summary statistics for the samples illustrated in fig. 28.2.

|  | Sample Size 10 | Sample Size 40 |
|---|---|---|
| mean of means | 500.7 | 499.2 |
| standard deviation of means | 24.92 | 13.29 |

This is the area to the right of 73" (after standardization) and we must subtract the table area from 1. From table A.1, TableArea$(1.43) = 0.92364$. Hence

$$\text{Prob}(x > 73) = 1 - \text{TableArea}(1.43) = 1 - .9236 = 0.0764$$

The probability that the height of a single man selected at random exceeds $6'1"$ is approximately 7.6%.

b) The corresponding $z$-value for a sample mean is

$$z = \frac{x - \mu}{\sigma/\sqrt{n}} = \frac{73 - 70}{2.1/\sqrt{25}} = 7.14$$

Since we are looking for the probability that the height is greater than $6'1"$. Since this is the area to the right of 73", we must subtract the table lookup area from 1. From table A.1, TableArea$(7.14) = 1.0000$. Hence

$$\text{Prob}(x > 73) = 1 - \text{TableArea}(7.14) = 1 - 1 = 0$$

The probability that the mean height of a sample of 25 randomly selected men exceeds $6'1"$ is essentially zero.

**Technology Example 28.2 (TI-84).** Repeat example 28.1 using a TI-84.

a) Use **normalcdf** to find the proportion to the right of 73 inches in an $N(70, 2.1)$ distribution.

    i. 2nd vars then scroll to **normalcdf**. Set **lower** to **73**, **upper** to a very large number, $\mu$ to 70, and $\sigma$ to 2.1, to produce:

$$\texttt{normalcdf(73, 1E99, 70, 2.1)}$$

    ii. The answer is **0.0765637714**.

b) Use **normalcdf** to find the proportion to the right of 73 inches in an $N(70, 2.1/\sqrt{25})$ distribution.

    i. 2nd vars then scroll to **normalcdf**. Set **lower** to **73**, **upper** to a very large number, $\mu$ to 70, and $\sigma$ to $2.1/\sqrt{25}$, to produce:

$$\texttt{normalcdf(73, 1E99, 70, 2.1/}\sqrt{25}\texttt{)}$$

    ii. The answer is **4.60244934E−13**, which is essentially zero.

**Technology Example 28.3 (R).** Repeat example 28.1 using R.

a) Use **pnorm** with **lower.tail=FALSE** to find the upper tail probability for (area to the right of) 73 inches in an $N(70, 2.1)$ distribution.

b) Use **pnorm** with **lower.tail=FALSE** to find the upper tail probability for (area to the right of) 73 inches in an $N(70, 2.1/\sqrt{25})$ distribution.

```
> pnorm(73,mean=70,sd=2.1,lower.tail=FALSE)
[1] 0.07656373
> pnorm(73,mean=70,sd=2.1/sqrt(25),lower.tail=FALSE)
[1] 4.570531e-13
```

# Exercises

1. Suppose $\mu = 30$, $\sigma = 5$. Find $P(x > 32)$. What is $P(\overline{x} > 32)$ if $n = 16$? if $n = 25$? (Ans: 34.46%, 5.48%, 2.28%)

2. Suppose $\mu = 17$, $\sigma = 3$. Find $P(x < 16)$. What is $P(\overline{x} < 16)$ if $n = 9$? if $n = 16$? (Ans: 36.94%, 9.12%, 4.78%)

3. Suppose $\mu = 10.5$, $\sigma = 2$. What is $P(8 < x < 10)$. What is $P(8 < \overline{x} < 10)$ if $n = 16$? if $n = 25$? (Ans: 29.56%, 15.97%, 10.57%)

4. Suppose a standardized test for seventh graders has an average score of 102 and a standard deviation of 12. What is the probability that a single student gets a score below 65? above 107? above 120? (Ans: 0.10%, 33.84%, 6.68%)

5. In question 4, what is the probability that the average score among a group of 12 students is above 107? in a group of 16 students? (Ans: 7.45%, 4.78%)

6. In question 4, what is the probability that the average score among a group of 25 students is between 96 and 112? What is the probability that a single student scores between 96 and 112? (Ans: 99.38%, 48.91%)

7. Suppose the year you take the Math SAT the average is 480 and the standard deviation is 95. What is the probability of scoring above 700? above 600? between 500 and 525? between 500 and 600? (Ans: 1.03%, 10.33%, 9.88%, 31.34%)

8. In question 7, suppose your high

school math class has 27 students who took the SAT What is the probability that the average score of all the students in the class was above 700? above 600? between 500 and 525? between 500 and 600? (Ans: 0, 0.000000000026, 13.00%, 13.7%)

# FAQ 29. How Do I Find $z^*$?

A confidence interval is defined with a specific level of confidence $C$, and that level of confidence defines a critical value $z^*$. For example, if you draw a random sample of size $n$ from a population that is distributed normally with standard deviation $\sigma$, then a confidence interval on the mean is given by

$$\mu = \overline{x} \pm \frac{z^* \sigma}{\sqrt{n}}$$

The confidence level $C$ is equal to the probability

$$C = P(-z^* < z < z^*)$$

as discussed in chapters 26 and 31.

The confidence level $C$ is sometimes given a decimal number less than one, such as 0.95, and is sometimes given as a percentage, as in 95%. A 95% confidence level means the same thing as $C = 0.95$, and the numbers are frequently written interchangeably, as in $C = 0.95 = 95\%$ (although your stats teacher might not like this notation, its done all the time) (fig 29.2).

## Critical Value

The **critical value** for a $z$ statistic with confidence level $C$ is the value $z^*$ such that the central area under the curve of the $N(0,1)$ (standard normal) distribution, between $-z^*$ and $z^*$, is equal to $C$.

Many textbooks have tables of the critical values tabulated as a function of confidence level $C$ (table 29.1). Sometimes these are combined with $t$-tables, as illustrated near the bottom of the table in appendix A.

Figure 29.1.: Illustration of critical values, as tabulated in table 29.1. The critical value for a confidence level $C$ is the number $z^*$ such that the central area under a normal distribution with mean 0 and standard deviation 1 between $-z^*$ and $z^*$ is equal to $C$. Each of the two tails have area

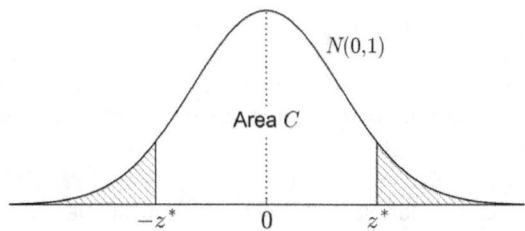

Figure 29.2.: Conversion of confidence level $C$ between units of proportion (values between 0 and 1) and percent (values between 0 and 100).

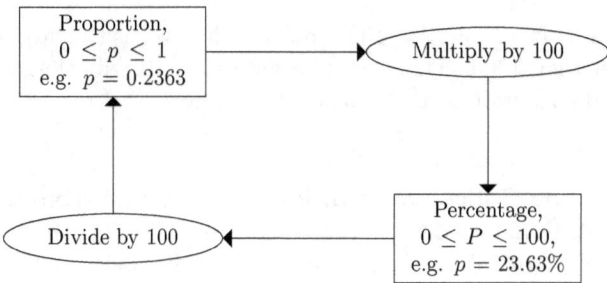

Table 29.1. Critical Values for the Normal Distribution. See figure 29.1 for definition of $C$ and $Z^*$.

| $C$ | 0.5 | 0.6 | 0.7 | 0.75 | 0.8 | 0.85 | 0.9 | 0.95 | 0.96 |
|---|---|---|---|---|---|---|---|---|---|
| $z^*$ | 0.6745 | 0.8416 | 1.0364 | 1.1503 | 1.2816 | 1.4395 | 1.6449 | 1.96 | 2.0537 |

| $C$ | 0.97 | 0.98 | 0.985 | 0.99 | 0.995 | 0.996 | 0.997 | 0.998 | 0.999 |
|---|---|---|---|---|---|---|---|---|---|
| $z^*$ | 2.1701 | 2.3263 | 2.4324 | 2.5758 | 2.807 | 2.8782 | 2.9677 | 3.0902 | 3.2905 |

To find the critical value $z^*$ in a two column table such as table 29.1:

1. Locate the desired confidence level $C$ in the first column.
2. Read the value of $z^*$ off from the the second column.

**Example 29.1.** Find the critical value corresponding to a confidence level of $C = 99.6\%$.

The critical value is $z^* = 2.782$, because 2.782 is written in the $z^*$ column next to the value of $C = 0.996$.

Sometimes tables of the t-distribution (ch. 37) list critical values for $z^*$. To find the critical value $z^*$ from a table of the $t$ distribution, such as in table A.2, we first check to see if there is a row labeled $z^*$. If there is such a row,

1. Locate the desired confidence level $C$ in the first row of the $t$-table.
2. Follow the column down to the row labeled $z^*$ and read off it's value from that row.

If there is no such row labeled $z^*$, look for a row labeled with a very large number of degrees of freedom. Typically $df = 1000$ will be sufficient for two decimal place accuracy. For example, for 95% confidence, $z^* = 1.960$, whereas $t^*_{1000} = 1.962$.

Use the value from the $df = 1000$ row, or whatever is the largest value if it is larger than 1000. This works because the $t$ distribution approaches a normal distribution as $df$ becomes very large.

**Estimating the Critical Value By Reverse Look-up from a Table of the Normal Distribution.** A reverse lookup (ch. ??) can also be used to estimate a critical value. Recall from the discussion above that the critical value is the area between $-z^*$ and $z^*$,

$$C = P(-z^* \leqslant z \leqslant z^*)$$

Since the total area under the curve is 1, the area of each tail $T$ is

$$T = \frac{1 - C}{2}$$

This formula allows us to do a reverse lookup for $z^*$ given $C$.

1. Calculate $T = (1 - C)/2$.

   **Example 29.2.** Find the critical value $z^*$ corresponding to $C = .9$ using table A.1.

   We first calculate the single-tail $z$ area,

   $$T = \frac{1 - .9}{2} = \frac{.1}{2} = .05$$

   (to be continued)

2. Reverse lookup the value of $z$ to get $z^*$ as (ch. 22).

(**Continuation of example 29.2.**) We look in table A.1 for the values closest to $T = 0.05$ (do a reverse lookup). The number 0.05 is bracketed by 0.495 and 0.505. Following the table to the row and column headers this tells us that $1.64 < z^* < 1.645$. Since $T$ is exactly halfway between the two numbers we can safely estimate that $z^*$ is also half way between:

$$z^* = \frac{1.64 + 1.65}{2} = 1.645$$

This agrees (to 3 decimal places) with the value in the table given above.

**Technology Example 29.3 (TI-84).** Find the critical value for a confidence level of $C = 0.97$ using a TI-84.

We can solve this problem using **invNorm** based on the single tail area.

1. The single tail area is $T = (1 - C)/2 = (1 - .97)/2 = 0.015$
2. Using 2nd vars, bring up the **DISTR** menu.
3. Scroll to **invNorm** and press enter.
4. Set **area** to **0.015**, $\mu$ to **0** and $\sigma$ to **1**.
5. Scroll to **Paste** and press enter. The command **invNorm(.015, 0, 1)** should be displayed on the screen. Press enter to get the answer.
6. The answer **−2.170090375** will be displayed. By symmetry, the critical values are $z^* = \pm 2.170090375$.

**Technology Example 29.4 (R).** Find the critical value for a confidence level of $C = 0.97$ using a R.

We can solve this problem using **qnorm** based on the single tail area.

```
> T=(1-.97)/2
> qnorm(T)
[1] -2.17009
```

By symmetry, the critical values are $\pm 2.17009$.

**Technology Example 29.5 (Spreadsheet).** Find the critical value for a confidence level of $C = 0.97$ using a spreadsheet.

The inverse of the standard normal is **norm.s.inv(z)**.

We can apply that to the left tail area, as we did in examples 29.3 and 29.4, by typing **=norm.s.inv((1−0.97)/2)** into any blank cell. The answer, **−2.1700904**, appears in the same cell.

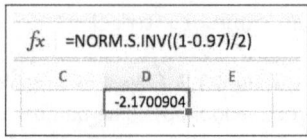

| $fx$ | =NORM.S.INV((1-0.97)/2) | |
|---|---|---|
| C | D | E |
| | -2.1700904 | |

# Exercises

Find the critical values for each of the following confidence levels to at least three decimal points.

1. $C = .95$ (ans: 1.960)

2. $C = .975$ (ans: 2.241)
3. $C = .94$ (ans: 1.881)
4. $C = .68$ (ans: 0.994)
5. $C = 90\%$ (ans: 1.645)
6. $C = 85\%$ (ans: 1.440)
7. $C = 92.5\%$ (ans: 1.780)
8. $C = 89\%$ (ans: 1.598)

# FAQ 30. How Do I Find a Confidence Interval On the Mean?

Suppose we draw random samples of size $n$ from a population that is normally distributed with known standard deviation $\sigma$. The distribution of the sample means will also be normally distributed (for sufficiently large $n$) with standard deviation $\sigma/\sqrt{n}$ (chapter 28). If we take a large number of samples, the level $C$ confidence interval gives us a range of values between which a proportion $C$ of the sample means are expected to fall. We use this interval as an estimate of the true population mean.

---

### Confidence Interval On The Mean

If a simple random sample of some variable $x$ is drawn from a population with known standard deviation $\sigma$ but unknown mean $\mu$, **a level $C$ confidence interval on the mean** $\mu$ is given by

$$\mu = \overline{x} \pm z^* \frac{\sigma}{\sqrt{n}}$$

where $z^*$ is the critical value (chapter 29), defined such that the central area between $-z^*$ and $z^*$, under the curve of a standard normal distribution, is equal to $C$.

---

The confidence level $C$ is equal to the probability

$$C = P(-z^* < z < z^*)$$

as discussed in chapters 26 and 31. It is sometimes given a decimal number less than one, such as 0.95, and is sometimes given as a percentage, as in 95%. A 95% confidence level means the same thing as $C = 0.95$, and the numbers are frequently written interchangeably, as in $C = 0.95 = 95\%$ (although your stats teacher might not like this notation, this is done all the time).

Figure 30.1.: The critical number $z^*$ for a level $C$ confidence interval is defined such that the central area under the curve of a normal distribution with mean zero and standard deviation one, between $z^*$ and $-z^*$, is equal to $C$.

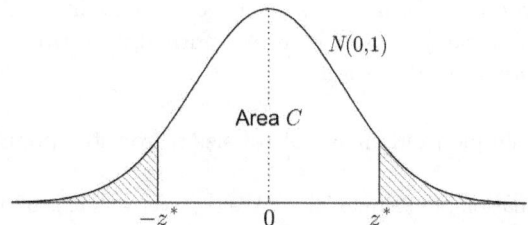

The term $z^*\sigma/\sqrt{n}$ is called the **margin of error**, as discussed in chapter 31. The margin of error is related to the confidence interval in the following manner:

$$(\text{confdence interval on } \underbrace{\text{parameter}}_{\text{e.g. } \mu}) = (\underbrace{\text{statistic}}_{\text{e.g. } \bar{x}}) \pm (\text{margin of error})$$

If the standard deviation of the population $\sigma$ is not known, the standard deviation of the sample may be used, and a $t$-test should be performed, instead of using the procedure here. This is discussed in chapter 41. To find a confidence interval using $z^*$, the standard deviation $\sigma$ must be known.

**Example 30.1.** Suppose you are an instructor grading a large number, say 225, exams. You used a well validated exam with grades typically distributed normally with $\sigma = 10$ points. To get an early estimate of the grades, pick a random sample of 15 exams to grade first adn determine a 95% on the mean. Here are the grades:

$$71, 66, 68, 60, 49, 60, 62, 69, 46, 51, 51, 59, 52, 65, 65$$

Figure 30.2.: Decision tree for using a z-procedure or a t-procedure. See also fig. 38.1

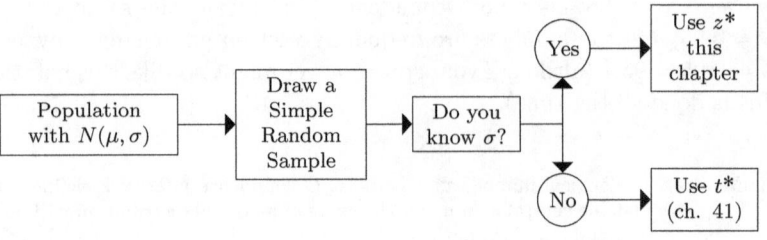

Here is the procedure for finding a level C confidence interval on the mean when you have a random sample from a normally distributed population with known standard deviation $\sigma$.

1. Take a simple random sample of size $n$ from the population.

   (**Continuation of example 30.1.**) We are given a sample of $n = 15$ values.

2. Calculate $\bar{x}$ as described in chapter 13.

   (**Continuation of example 30.1.**) The mean of the given sample is $\bar{x} = 59.60$.

3. Verify that you know the standard deviation of the population, $\sigma$. If you do not know $\sigma$, you must perform a $t$ procedure instead, as described in chapter 41.

   (**Continuation of example 30.1.**) We are told that the population has a standard deviation of $\sigma = 10$.

4. Determine the critical value $z^*$ corresponding to the confidence level $C$ using either technology (e.g., a calculator function or button) or a table (such as table 29.1 or table A.2 (see chapter 29).

   (**Continuation of example 30.1.**) From table 29.1, for $C = 95\%$, the critical value is $z^* = 1.960$.

5. Calculate the margin of error (m.e.) using the formula

$$(\text{m.e.}) = z^* \frac{\sigma}{\sqrt{n}}$$

(**Continuation of example 30.1.**) We know from step 1 that $n = 15$; from step 3 that $\sigma = 10$; and from step 4 that $z^* = 1.960$. Therefore

$$\text{(m.e.)} = (1.960)\frac{10}{\sqrt{15}} = 5.06$$

6. Calculate the confidence interval using

$$\mu = \bar{x} \pm \text{(m.e.)}$$

or in interval notation

$$\bar{x} - \text{(m.e.)} \leqslant \mu \leqslant \bar{x} + \text{(m.e.)}$$

(**Continuation of example 30.1.**) From item 2, we know that $\bar{x} = 59.6$. Therefore the 95% confidence interval is

$$59.6 - 5.06 \leqslant \mu \leqslant 59.6 + 5.06$$

. Simplifying gives

$$54.54 \leqslant \mu \leqslant 64.66$$

**Technology Example 30.2 (TI-84).** Repeat example 30.1 using a TI-84. We use the **ZInterval** function for this.

1. Press $\boxed{\text{stat}}$ then scroll over to $\boxed{\text{TESTS}} \gg \boxed{\text{ZInterval}}$ $\boxed{\text{enter}}$, or $\boxed{\text{tests}}$ $\boxed{7}$
2. For **Inpt** select **Stats**; the set $\sigma$ to 10, $\bar{x}$ to 59.6, $n$ to 15, and **C–Level** to 0.95.
3. Scroll to **Calculate** and press $\boxed{\text{enter}}$. The interval **(54.539, 64.661)** will be displayed. This is the solution (compare with example 30.1).

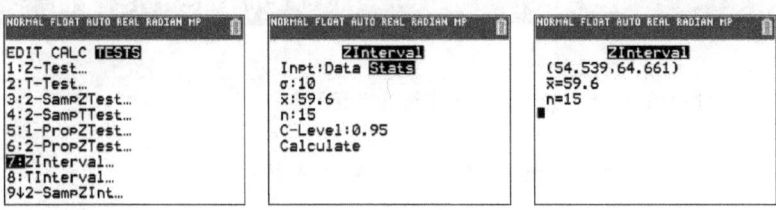

**Technology Example 30.3 (R).** Repeat example 30.1 in R.

We can use **qnorm** to determine the critical value $z^*$, and then calculate the margin of error as $ME = z^*\sigma/\sqrt{n}$.

```
> n=15
> sigma=10
> xbar=59.6
> C=0.95
> z=qnorm((1-C)/2,lower.tail=FALSE)
> ME = z*sigma/sqrt(n)
> xbar-ME
[1] 54.53939
> xbar+ME
[1] 64.66061
```

Thus the confidence interval is from 54.54939 to 64.66061.

## Exercises

1. The verbal SAT is normally distributed with $\sigma = 100$. What is the margin of error for a 95% confidence level when the sample size is 50? when the sample size is 100? when the sample size = 200? (Ans: 27.7, 19.6, 13.9)

2. Repeat the previous problem for a 99% confidence interval. (Ans: 36.4, 25.8, 18.2)

3. The following random sample is drawn from SAT verbal test scores, which are normally distributed with $\sigma = 100$:
676, 495, 577, 479, 640, 396, 588, 420, 421, 599, 621, 490, 613, 526, 522, 361
Find a 97% confidence interval on the mean. (Ans: $z^* = 2.17$; $n = 16$; $\bar{x} = 526.5$; $472.25 < x < 580.75$)

4. Assume the sample size was 115 in the previous problem, with $\bar{x} = 526.5$. What would the 97% confidence interval be? (Ans: $506.26 < x < 546.74$)

## FAQ 31. How Big Does My Sample Size Have to Be to Do Inference of the Mean?

Recall that a confidence interval is defined as

$$\text{parameter} = \text{estimate} \pm (\text{margin of error})$$

If we collect a simple random sample of size $n$ from a normal population of known standard deviation $\sigma$ then a level C confidence interval on the mean (chapter 30) is given by

$$\mu = \bar{x} \pm z^* \frac{\sigma}{\sqrt{n}}$$

The term $z^*\sigma/\sqrt{n}$ is called the **Margin of Error**. It is half the width of the confidence interval. It measures the distance from the calculated value of $\bar{x}$ to the edge of the interval. The critical value $z^*$ is determined based on the desired level of confidence: $C$ is the central area in a normal distribution with mean 0 and standard deviation 1 between $-z^*$ and $z^*$.

To find the minimum sample size that will produced the desired margin of error for a given confidence interval, we plug the desired values of $M$, $z^*$, and $\sigma$ into the formula and solve for $n$. Setting $M$ equal to the formula for the margin of error gives

$$M = z^* \frac{\sigma}{\sqrt{n}}$$

Multiplying both sides of the equation by $\sqrt{n}/M$

$$\frac{\sqrt{n}}{M} \times M = z^* \frac{\sigma}{\sqrt{n}} \times \frac{\sqrt{n}}{M}$$

The $\sqrt{n}$'s cancel on the right, and the $M$'s cancel on the left,

$$\sqrt{n} = \frac{z^*\sigma}{M}$$

Squaring both sides of the equation gives the minimum sample size:

$$n = \left[\frac{z^*\sigma}{M}\right]^2$$

Since $n$ refers to a number of individuals, the result must be rounded up to the next largest integer.

---

### Minimum Sample Size

The minimum sample size to give a margin of error $M$ with confidence level $C$ in a normally distributed population with with standard deviation $\sigma$ is

$$n = \left[\frac{z^*\sigma}{M}\right]^2$$

where $z^*$ is the critical value such that the central area under the normal curve between $-z^*$ and $z^*$ is $C$.

---

**Example 31.1.** The standard deviation on the Graduate Record Exam (GRE) is 93 points. Suppose you want to take a sample of students at your university to estimate the mean score to within 10 points with 95% confidence. How many students would you have to sample? (to be continued)

To find the minimum sample size $n$ required to get a specified margin of error $M$ at a confidence level $C$:

1. Find $z^*$ based on your confidence level $C$.

    **(Continuation of example 31.1.)**
    For $C = 95\%$ we get $z^* = 1.960$ (table A.2).

2. Determine $M$ (the desired margin of error) from the problem. statement This is always in raw data units, i.e. the same units as $\mu$ or $\bar{x}$.

    **(Continuation of example 31.1.)**
    We are asked for a margin of error of 10 points, i.e., $M = 10$.

3. Determine the population standard deviation $\sigma$ from the problem domain, i.e., information about the problem. This information will come from a domain expert, scientist, reference book, etc.
    **(Continuation of example 31.1.)** We are given that $\sigma = 93$.

4. Calculate $n$ using the formula $n = \left[ \dfrac{z^* \sigma}{M} \right]^2$.

    **(Continuation of example 31.1.)**
    Combining the last three steps,

    $$ n = \left[ \frac{(1.960)(93)}{10} \right]^2 = 18.228^2 = 332.26 $$

5. Since the answer must be an integer, round up. The number must be an integer because we are estimating the number of *individuals* to sample, and you can't sample a fraction of an individual. The formula gives the *minimum* value required, so we must round *up*, and not down.

    **(Continuation of example 31.1.)**
    For our data, rounding up gives $n = 333$.

# Exercises

1. Suppose the GRE has a standard deviation of 100. How large a sample is required to estimate the mean GRE to within 10 points at 99% confidence? To within 99.5% confidence?
2. Repeat the previous problem but estimate to within 5 points.
3. The height of American men ages 20-29 is normally distributed with a standard deviation of 5 cm. How large of a sample is required to estimate the mean height of American men ages 20-29 with 99% confidence to within 2 cm?
4. The Smello milk company claims that their milk continues to smell fresh for at least ten days after the expiration date with a standard deviation of 1.2 days. How many cartons of milk would you have to smell to estimate the actual smell-fresh period to within 0.25 days with a 98% confidence?

# FAQ 32. What is a Hypothesis?

### Hypothesis

A **hypothesis** is a statement about a parameter.

**A hypothesis never mentions a statistic** (a number that describes a sample like $\bar{x}$), **only a parameter** (a number that describes the population, like $\mu$) (ch. 4).

Statisticians talk about two types of hypotheses: the **Null Hypothesis** and an **Alternative Hypothesis**.

The **Null Hypothesis** is some assumption that is assumed to be true until proven otherwise. When we perform a hypothesis test (ch. 35) we assume that the null hypothesis is true. If the evidence is significant, then we reject the null hypothesis and replace it with something else (e.g., the alternative hypothesis).

### Significant

The result of a hypothesis test is said to be **significant** if it provides sufficiently strong evidence to support the alternative hypothesis and reject the null hypothesis.

We label the null hypothesis as $H_0$ and state it mathematically as an equality. If it is a statement about the mean, for example, and $\mu_0$ is some fixed constant number, then we write the Null Hypothesis as:

$$H_0 : \mu = \mu_0$$

**Example 32.1.** State the null hypothesis that the average height all men in the United States is 72 inches.

A hypothesis is a statement about a parameter. In this case, the parameter is the average, or mean, so it refers to the population mean $\mu$. The Null Hypothesis then becomes $H_0 : \mu = 72$.

When we believe that there is reason to doubt the Null Hypothesis, we state an **Alternate Hypothesis**. An Alternate Hypothesis is denoted by the symbol $H_A$, $H_a$, or $H_1$. Alternate Hypotheses are expressed as inequalities ($\mu > \mu_0$, $\mu < \mu_0$, or $\mu \neq \mu_0$, and are represented using the same symbols and values mentioned in the Null Hypothesis. If the Null Hypothesis is $H_0 : \mu = \mu_0$, then there are three possible Alternate Hypotheses (fig. 32.1). The first two are called **one-sided hypotheses** because we believe the correct value is on one side of $\mu_0$, but not the other. The third is called a **two-sided hypothesis** because we believe $H_0$ is wrong, but we do not know on which side the correct value occurs.

1. A one sided hypothesis: $H_A : \mu < \mu_0$.
2. A one sided hypothesis: $H_A : \mu > \mu_0$.
3. A two sided hypothesis: $H_A : \mu \neq \mu_0$.

Figure 32.1.: Classification of alternate hypotheses.

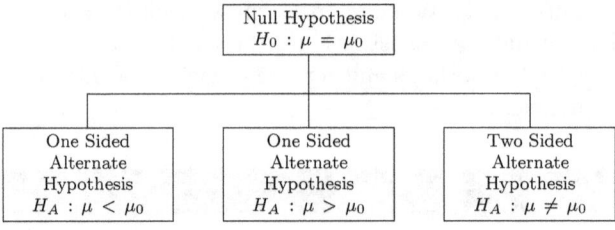

**Example 32.2.** Suppose you believe that that the average height of all American men is less than 72 inches. Using the null hypothesis from example 32.1, state the Null and Alternate hypotheses.

From example 32.1 the null hypothesis is $H_0: \mu = 72$. Since we believe the true height is smaller than 72, we make the one sided Alternate Hypothesis $H_A :: \mu < 72$.

**Example 32.3.** Suppose we know that the average height of boys age 12 in some population is 149 centimeters, but we do not know the height of girls in our population. We believe that the height of girls is different from that of boys. Make appropriate hypotheses.

We make the hypotheses $H_0: \mu = 149$ and $H_A: \mu \neq 149$. We use the two sided hypothesis because we guess that girls heights are different, but we are not sure in which direction.

# Exercises

Formulate null and alternate hypotheses for each of the following.

1. A popular TV ad that says you can't eat just one.
2. A friend of yours says that McBurgers' Big Fries contain precisely 43 French Fries per package. You believe there are more.
3. A candy company says that every package of Emm-Enn-Emm candies contains exactly 50 colorfully coated chocolate candies. You suspect that there are fewer.
4. Your stats professor says that the failure rate in introductory statistics is 45%. You heard elsewhere that it is higher.
5. Your friend says that the postage rate is 50 cents per ounce. You think that is not right but you are not sure what it is

# FAQ 33. What is a p-value?

P-values are used to evaluate hypothesis tests (see chapter 35). In this chapter we will discuss the p-value for a **hypothesis test on the mean of a population when the distribution is normal and the standard deviation $\sigma$ of the population is known.** In this type of hypothesis test, we would collect a random sample of size $n$ and calculate its mean value $\bar{x}$. We would make a null hypothesis $H_0$, typically of the form $H_0 : \mu = \mu_0$ (here $\mu_0$ is a known, fixed constant). We would then asked a question that can formulated in terms of an alternate hypothesis (ch. 32). The alternate hypothesis can be stated in one of three forms (fig. 32.1).

The p-value is formulated as follows. Draw a simple random sample and calculate $\bar{x} = x_0$. The p-value is the probability of $\bar{x}$ being at least as far from the population mean $\mu$ as $x_0$, assuming that the null hypothesis is true. Since $\mu = \mu_0$ when $H_0$ is true, this statement can be rewritten as: The p-value is the probability that $\bar{x}$ is at least as far from $\mu_0$ as $x_0$.

---

**p value**

The **p value** of a statistic $x_0$ is the probability of an observation of $x$ being at least as far from it's expected value[a] as the observed value of the statistic $x_0$, assuming that the null hypothesis is true.

---

[a]The expected value is the mean of the population.

---

If we normalize to $z = (x_0 - \mu_0)/(\sigma/\sqrt{n})$, then the p-value is the probility that $Z$ is at least as far away from 0 as $z$ is. This p-value can be found from a table of the standard normal distribution (table A.1), but the method for converting the table entry is different for each type of alternate hypothesis (fig. 33.1).

Figure 33.1.: Calculation of p-values for different types of alternate hypotheses. Notation: $p$: p-value; $P$: probability; Area: area by table lookup, e.g., in table A.1.

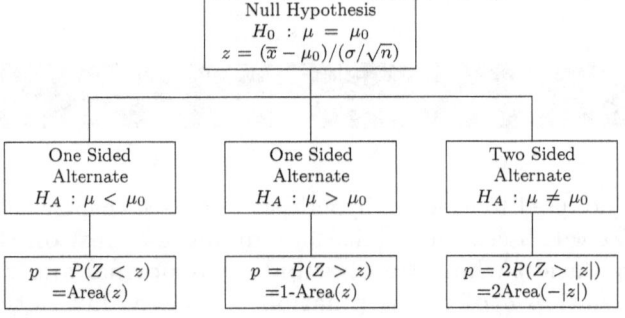

To test a hypothesis[1] $H_0 : \mu = \mu_0$ we obtain a random sample of size $n$. Let $\sigma$ be the population mean. Then,

1. Calculate the $z$−score $z = \dfrac{\bar{x} - \mu_0}{\sigma/\sqrt{n}}$.
2. The case when $H_A : \mu < \mu_0$ is illustrated in figure 33.2. ( If the

---

[1]Assuming that the population is randomly distributed with standard deviation $\sigma$ and that we can draw an SRS.

alternate hypothesis is not: $H_A : \mu < \mu_0$, skip to step 3. ) In this case, the probability that an $\bar{x}$ will be observed, that is as far away from $\mu_0$, or further away from $\mu_0$, as the one actually observed, **assuming that $H_0$ is true**, is given by the area under the curve of the distribution and **to the left** of $z$. This is the shaded area in fig. 33.2. Thus $p =$TableArea$(z)$, i.e., $p$ can be found by a direct lookup into a table of the normal distribution, such as into table A.1.

**Example 33.1.** A particular type of cable needs to be able sustain a current of 3 Amperes in order to charge a cell phone safely. The maximum sustainable current can be tested with a Magico cable testing device. The Magico Cable company tests a sample of 25 of its cables and determines that the average safe charging current is 2.9 Amperes. Using a standard deviation of 0.25 Amperes, what is the p-value for this test?

We make the null hypothesis $H_0 : \mu = 3$ and alternative hypothesis $H_A : \mu < 3$. The z-score is $z = (\bar{x} - \mu_0)/(\sigma/\sqrt{n}) = (2.9 - 3)/(.25/\sqrt{25}) = -.1/.05 = 2$. The p-value is $P(z < -2) =$TableArea$(-2)=0.0228$.

We do not make a conclusion about whether or not this p-value is significant because we did not declare an $\alpha$ value (chapter 34). In order to determine whether or not the test is significant, we must compare whether or not $p < \alpha$.

Note that some statisticians would disagree, and would define $\alpha$ after the fact. They would say "this result is significant at the 2.28% level."

**Technology Example 33.2 (TI-84).** Repeat example 33.1 using the TI-84. We use the **Z-Test** function for $\mu < \mu_0$.

a) Press stat then select TESTS $\gg$ Z-Test enter, or TESTS 1 .

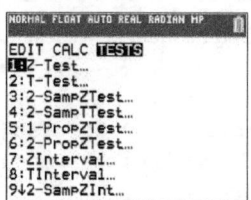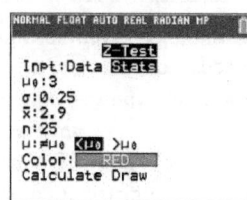

b) On the **Z-Test** menu set: **Inpt** to **Stats**; $\mu_0$ to 3; $\sigma$ to 0.25; $\bar{x}$ to 2.9; $n$ to 25; and select $\mu < \mu_0$.
c) Select **Calculate** and press enter. The $p$ value (0.022750062) and $z$ score (-2) for the test appear on the screen.

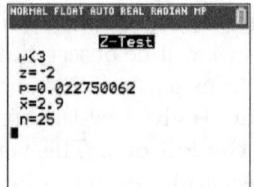 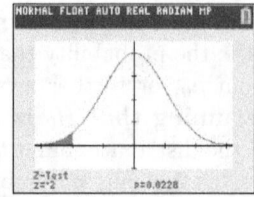

d) To plot the area corresponding by this test on a normal distribution, select **Draw** in step 2a instead of **Calculate**.

**Technology Example 33.3 (R).** Repeat example 33.1 in R.

We can emulate the calculations in example 33.1 step by step and get the p-value using **pnorm**.

```
> mu0=3
> sigma=0.25
> xbar=2.9
> n=25
> z=(xbar-mu0)/(sigma/sqrt(n))
> pnorm(z)
[1] 0.02275013
```

We can also skip the intermediate steps and type the entire expression in a single line.

```
> pnorm((2.9-3)/(.25/sqrt(25)))
[1] 0.02275013
```

The easiest way is to use **mean** and **sd** parameters for **pnorm**, though we still have to remember to divide the standard deviation by $\sqrt{n}$:

```
> pnorm(2.9, mean=3, sd=.25/sqrt(25))
[1] 0.02275013
```

**Technology Example 33.4 (Spreadsheet).** Repeat example 33.1 using a spreadsheet.

In any cell type the expression:

$$\texttt{=norm.dist(2.9,3,0.25/dist(25),1)}$$

followed by enter. The **norm.dist** function gives the area under a curve of a normal distribution; the arguments are:

$$\texttt{norm.dist(x, mean, sd, cum)}$$

where **cum** is set to 1 to indicate a cumulative distribution. We divide the standard deviation by $\sqrt{n}$ because the distribution represents the sample means. The answer (the p value) **0.022750132** appears in the cell after you hit enter:

| $fx$ | =NORM.DIST(2.9,3,0.25/SQRT(25),1) |
|---|---|

| C | D | E | F |
|---|---|---|---|
| 0.02275013 | | | |

Figure 33.2.: Definition of p-value when $H_A : \mu < \mu_0$. The p-value is the area under the normal curve to the left of $(\bar{x} - \mu_0)/(\sigma/\sqrt{n})$. This is precisely the probability that a vallue will be as far from $\mu_0$ or further from $\mu_0$, assuming that $H_0$ is true.

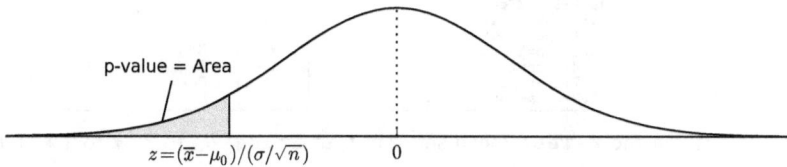

$z = (\bar{x} - \mu_0)/(\sigma/\sqrt{n})$      0

3. The case when $H_A : \mu > \mu_0$ is illustrated in figure 33.3. ( If the alternate hypothesis is not: $H_A : \mu > \mu_0$, skip to step 4. ) In this case, the probability that an $\bar{x}$ will be observed, that is as far away from $\mu_0$, or further away from $\mu_0$, as the one actually observed, **assuming that $H_0$ is true**, is given by the area under the curve of the distribution and **to the right** of $z$. This is the shaded area in fig. 33.3. Thus $p = 1 - \text{TableArea}(z)$, i.e., $p$ can be found by a direct lookup into a table of the normal distribution, such as into table A.1, and subtracting the result from one.

**Example 33.5.** You visit a college that you are considering attending and are told that students don't have to do more than 3 hours of homework per week, with a standard deviation of 1.8 hours. You walk around campus see a lot of students who look really tired so you suspect that they are really working a lot harder. You start asking students how much homeork they do, and after you have asked 36 students you determine that the students you interviewed do an average of 3.5 hours of homework per week. What is the p-value?

The null hypothesis is $H_0 : \mu = 3$. Since you think the students are doing more homework than 3 hours, the alternate hypothesis is $H_A : \mu > 3$.

The z-score is $z = (\bar{x} - \mu_0)/(\sigma/\sqrt{n}) = (3.5 - 3)/(1.8/\sqrt{36}) = 0.5/.3 = 1.67$

The alternate hypothesis refers to a right tail probability (see 33.1), p-value

---

$= P(z > 1.67) = 1 - P(z < 1.67) = 1 - \text{TableArea}(1.67) = 1 - .9525 = .0475$.
Thus the p-value is 4.75%. In terms of strict statistics, we would stop here because we do not have an $\alpha$. In terms of more naive analysis we would say that the test is significant at the 4.75% level.

**Technology Example 33.6 (TI-84).** Repeat example 33.5 using the TI-84. We use the **Z-Test** function for $\mu > \mu_0$.

a) Press stat then select TESTS ⟩ Z-Test enter, or TESTS 1 .

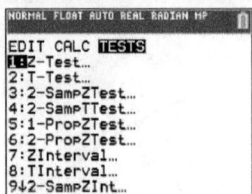 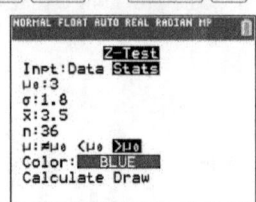

b) On the **Z-Test** menu set: **Inpt** to **Stats**; $\mu_0$ to 3; $\sigma$ to 1.8; $\bar{x}$ to 3.5; $n$ to 36; and select $\mu > \mu_0$.

c) Select **Calculate** and press enter . The $p$ value (0.0477903304) and $z$ score (1.66666667) for the test appear on the screen.

 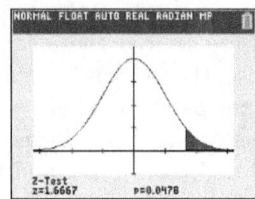

d) To plot the area corresponding by this test on a normal distribution, select **Draw** in step 3a instead of **Calculate**.

**Technology Example 33.7 (R).** Repeat example 33.5 in R.

Use **pnorm** with the **mean** and **sd** parameters, remembering to divide $\sigma$ by $\sqrt{n}$. We also need to set **lower.tail=FALSE** to indicate that we are doing an test for $\mu > \mu_0$ (i.e., finding the area to the right of $\mu_0$).

```
> pnorm(3.5, mean=3, sd=1.8/sqrt(36),
    lower.tail=FALSE)
[1] 0.04779035
```

**Technology Example 33.8 (Spreadsheet).** Repeat example 33.5 using a spreadsheet.

In any cell type the expression:

$$=1-\text{NORM.DIST}(3.5, 3, 1.8/\text{SQRT}(36),1)$$

followed by enter . The **norm.dist** function gives the area under a curve of a normal distribution; the arguments are:

$$\textbf{norm.dist}(\textbf{x, mean, sd, cum})$$

where **cum** is set to 1 to indicate a cumulative distribution. We divide the standard deviation by $\sqrt{n}$ because the distribution represents the sample means. We subtract from one because we are finding a right-tail area rather than a left-tail area. The answer (the p value) **0.047790352** appears in the cell after you hit ⎡enter⎤:

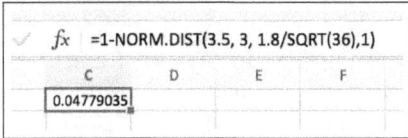

Figure 33.3.: Definition of p-value when $H_A : \mu > \mu_0$. The p-value is the area under a standard normal distribution to the right of $(\bar{x} - \mu_0)/(\sigma/\sqrt{n})$. This is precisely the probability that the sample mean is at least as far from $\mu_0$ as the observed value.

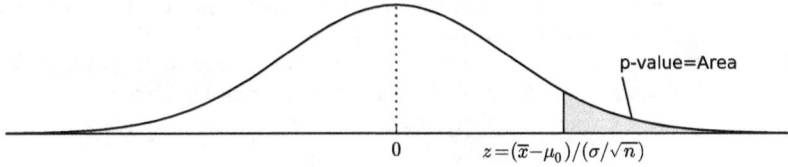

4. The only remaining possibility is that $H_A : \mu \neq \mu_0$. In this case, the probability that an $\bar{x}$ will be observed, that is as far away from $\mu_0$, or further away from $\mu_0$, as the one actually observed, **assuming that $H_0$ is true**, is given by the area under the curve of the distribution and in each of the tails, **to the right** of $|z|$ and **to the left** of $-|z|$. The two tails are shaded in fig. 33.4. Since the areas of the two tails are equal by symmetry, we only need to calculate the area of one and then double it. We can find the left tail area by direct table lookkup into table A.1. Thus $p = 2 \times \text{TableArea}(-|z|)$.

**Example 33.9.** Big Bin Company sells food by the cup out of big bins. There is a sign next to a bin of chocolates that says that one cup will contain about 23 candies. You suspect that this is not correct. (a) Formulate this as a two-sided hypothesis. (b) Suppose that after collecting 11 cups of candy and counting the amount in each cup you compute $\bar{x} = 23.27$. Find the p-value, assuming that $\sigma = 1$.

a) The null hypothesis is $H_0 : \mu = 23$. We suspect that is wrong, but we don't know which way, so our alternative hypothesis is $H_A : \mu \neq 23$.

b) The z-score is given by $z = (\bar{x} - \mu_0)/(\sigma/\sqrt{n}) = (23.27 - 23)/(1/\sqrt{11}) = .27/.30 = .90$.

Consequently the p-value comes out to be $2 \times \text{TableArea}(-|z|) = 2 \times \text{TableArea}(-.90) = 2(.1841) = .3682$, or 36.82%.

**Technology Example 33.10 (TI-84).** Repeat example 33.9 using a TI-84
We use the **Z-Test** function for $\mu \neq \mu_0$.

a) Press stat then select TESTS ⟩ Z-Test enter, or TESTS 1 .

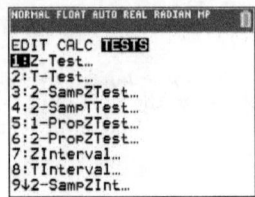 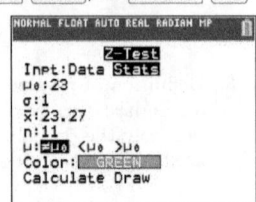

b) On the **Z-Test** menu set: **Inpt** to **Stats**; $\mu_0$ to 23; $\sigma$ to 1; $\bar{x}$ to 23.27; $n$ to 11; and select $\mu \neq \mu_0$.
c) Select **Calculate** and press enter. The $p$ value (0.3705258361) and $z$ score (0.8954886934) for the test appear on the screen.

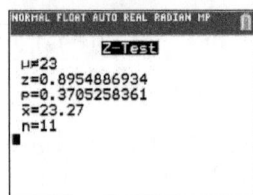 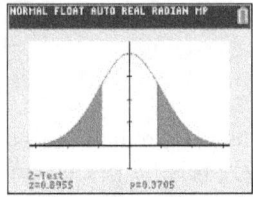

d) To plot the area corresponding by this test on a normal distribution, select **Draw** in step 3a instead of **Calculate**.

**Technology Example 33.11 (R).** Repeat example 33.9 using R.

Use the fact that the p-value for a 2-sided test is $2P(Z > |z|)$. In **pnorm**, set the parameters **mean=23**; **sd=1/sqrt (11)** (which means $1/\sqrt{11}$); and **lower.tail** to **FALSE**.

```
> 2*pnorm(23.27, mean=23, sd = 1/sqrt(11),
    lower.tail=FALSE)
[1] 0.3705259
```

**Technology Example 33.12 (Spreadsheet).** Repeat example 33.9 using a spreadsheet.

FAQ 33. What is a p-value?

Figure 33.4.: Definition of p-value for $H_A : \mu \neq \mu_0$. The p-value is equal to the sum of the two tail areas under the curve of a standard normal distribution, to the right of $|\bar{x} - \mu_0|/(\sigma/\sqrt{n})$ and to the left of $-|\bar{x} - \mu_0|/(\sigma/\sqrt{n})$. This is precisely the probability that the sample mean would take on a value as far from $mu_0$ as the value that was actually observed, or further away from $\mu_0$, and in either direction.

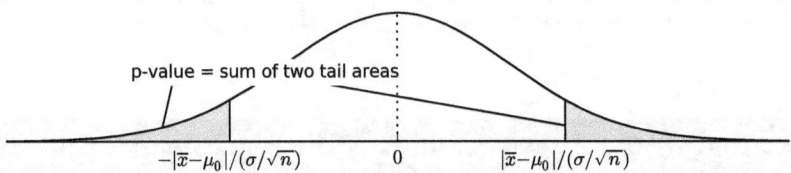

In any cell type the expression:

$$=2*(1-\texttt{NORM.DIST}(23.27, 23, 1/\texttt{SQRT}(11),1))$$

followed by enter. The **norm.dist** function gives the area under a curve of a normal distribution; the arguments are:

$$\texttt{norm.dist(x, mean, sd, cum)}$$

where **cum** is set to 1 to indicate a cumulative distribution. We divide the standard deviation by $\sqrt{n}$ because the distribution represents the sample means. We subtract from one because we are finding a right-tail area rather than a left-tail area. This result is doubled because we are performing a two tailed test. The formula thus gives $2P(Z > |z|)$. The answer (the p value) **0.370525907** appears in the cell after you hit enter:

| *fx* | =2*(1-NORM.DIST(23.27, 23, 1/SQRT(11), 1)) | | | |
|---|---|---|---|---|
| C | D | E | F | G |
| 0.37052591 | | | | |

# Exercises

For each of the following (a) formulate both null and alternative hypothesis $H_0$ and $H_A$; and (b) calculate the p-value.

1. A populate TV ad says you can't eat just one. You buy several bags of chips and ask 15 friends to "have a chip." You count how many chips they eat to determine $\bar{x} = 7.2$. Assume $\sigma = 5$

2. A friend of yours says that McBurgers' Big Fries contain precisely 43 French Fries per package. You believe there are more, so you order six bags of fries and count each one. You find $\bar{x} = 45$. Assue $\sigma = 3$.

3. A candy company says that every package of Emm-Enn-Emm candies contains exactly 50 colorfully coated chocolate candies. You suspect that there are fewer, so you buy 17 bags of candy and count

the numy of chocolates in each bag. You determine that the average number of candies is 48.5. Assume $\sigma = 5$.

4. Your stats professor says that the failure rate in introductory statistics is 45%. You heard elsewhere that it is higher. You file a Freedom of Information Act request with the university to get a sample of the course passage results over the past the 10 years. They tell you the following: that based on a random sample of 73 randomly selected students, the pass rate was 53%. Assume $\sigma = 3\%$. Assume that the failure rate is 100 minus the pass rate.

# FAQ 34. How Small is Small? What is alpha? What is the Difference Between p and alpha?

The p-value, in and of itself, does not tell us what to conclude in a hypothesis test. It is nothing more than a mathematical statement of probability under the assumption that $H_0 : \mu = \mu_0$ is true.

Since the p-value gives the probability that a measurement is as far from $\mu_0$, or further, then we actually observed, assuming that $H_0$ true, it is really a measurement of the rarity of our observation. The further away our measurement is, the rarer it becomes.

**All that the p-value tells us is how rare** (or how possibly, how common) **our observation actually is**, assuming that the null hypothesis is true.

---

**What Does the P-value really say?**

**The p-value tells us how rare an observation is**, assuming $H_0$ is true.

---

**Example 34.1.** Suppose we calculate a p-value of 0.04. What does this really mean?

Observations of what we saw occur with a rarity of $p = 0.04$. Since $1/.04 = 25$, this means that observations like the one we observed may occur by random chance as often as one in 25 times if the experiment or survey is repeated many times assuming that the null hypothesis is true.

**Example 34.2.** What does a p-value of $0.14 mean$? a p-value of $0.001$?

Since $1/0.14 \approx 7$, a p-value of $0.14$ means that measurements as rare as ours may occur approximately once in seven experiments, assuming that the null hypothesis is true.

Since $1/.001 = 1000$, a p-value of $0.001$ means that measurements as rare as ours may occur approximately once in a thousand experiments, assuming that the null hypothesis is true.

We don't want to jump to a conclusion that we observed something really important if what we saw might have occurred by random chance.

---

### p-value and Random Chance

The p-value is probability that an event might have occurred by random chance, assuming that the null hypothesis is true.

---

If the event (observation) is extremely rare (low p-value) then it is extremely unlikely to have occurred by random chance when the null hypothesis is true. This means we want to pay very close attention to such observations.

If the event is not very rare (not a low p-value) that it might very well have occured by random chance when the null hypothesis is true. In this kind of situation we would expect such observations to be very meaningful and would not want to jump to any conclusions.

In order to distinguish to between the two situations of "low p-value" and "not a low p-value" we need to define a cutoff. This cutoff is designated by the Greek letter $\alpha$ (alpha).

---

### How Small is Small

The number $\alpha$ (alpha) defines a political (not a mathematical) boundary for statistical significance. If $p \leq \alpha$, we say that the observation is statistically significant. If $p > \alpha$, it is not.

---

A hypothesis test (ch. 35) is said to be **statistically significant at a level** $\alpha$ if (table 34.1)

▸ The observed results are unlikely to occur by random chance, assuming the null hypothesis $H_0$ is true; and
▸ The probability $p$ (the $p$-value) that the observed results do occur is smaller than or equal to $\alpha$ (i.e., $p \leqslant \alpha$).

This means that **the significance of statistical result depends on the value of** $\alpha$.

---

**Significance and alpha**

A result that is valid at one level of $\alpha$ may not be significant at another level.

---

Table 34.1.: Assuming the null hypothesis $H_0$ is true, we decide wither or not to accept $H_A$ and reject $H_0$ by comparing $p$ to $\alpha$. If $p \leqslant \alpha$, the event is rare and significant; otherwise it is not. We accept $H_A$ if the event is significant.

| Event | Compare | Rare | Significant | Accept | Reject |
|---|---|---|---|---|---|
| Probability | $p$ with $\alpha$ | | | $H_A$ | $H_0$ |
| p is small | $p \leqslant \alpha$ | Yes | Yes | Yes | Yes |
| p is large | $p > \alpha$ | No | No | No | No |

**Example 34.3.** The SAT typically has a mean of 500 and standard deviation of 100. You suspect the average is higher, and collect a random sample of 125 scores. You find that $\bar{x} = 515$. Is this significant at an level of $\alpha = 5\%$? at $\alpha = 1\%$?

The procedure for hypothesis tests is given in chapter 35. We use the hypotheses

$$H_0 : \mu = 500$$
$$H_A : \mu > 500$$

The z-value is

$$z = \frac{\bar{x} - \mu_0}{\sigma/\sqrt{n}} = \frac{515 - 500}{100/\sqrt{125}} = 1.677$$

From table A.1 we get $p = 0.047$.

▸ If we are testing with $\alpha = 5\%$, then this is evidence in support of $H_A$ because $.047 < 0.05$. We would reject the the null hypothesis, because this would be considered a sufficiently rare event under the assumption that $H_0$ is true.

---

FAQ 34. How Small is Small?

▸ If we were testing with $\alpha = 0.01$, then we would observe that $p = 0.047 > 0.01 = \alpha$ and therefore this event is likely to occur to by random chance under the assumption that $H_0$ is true. In this case, we would not consider this observation evidence in support of the alternative hypothesis and we would not reject the null hypothesis.

There is no "right" answer in this example. It depends entirely on what you believe what is a rare event. Most studies use $\alpha = 0.05$ or smaller. If we choose $\alpha = 0.05$, then we will obtain evidence against $H_0$ by random chance one time out of 20. If we choose $\alpha = 0.01$, then this will only occur one time out of 100.

**One Sided Alternate Hypothesis** $H_A: \mu > \mu_0$. The situation when the one-sided alternative hypothesis $H_A: \mu > mu_0$ is tested against the null hypothesis $H_0: \mu = \mu_0$ at a confidence level $\alpha$ is illustrated in fig 34.1

Figure 34.1.: Accept and reject region for $H_A: \mu > \mu_0$. The null hypothesis is rejected if the normalized observation falls into the shaded area.

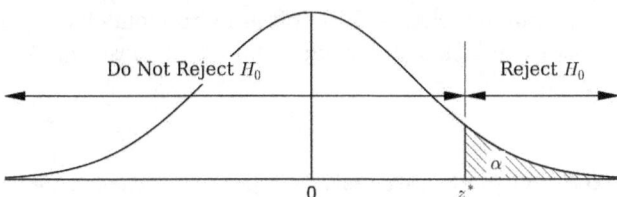

The curve shows the standard normal curve assuming that $H_0$ is true. The critical value $z^*$ corresponds to $\alpha$. The shaded region has area $\alpha$, and it gives the probability

$$\alpha = P(z \geqslant z^*)$$

Changing the size of the tail region is equivalent to changing the value of $\alpha$. Smaller values of $\alpha$ will lead to more convincing results.

Your statistics will allow you to calculate a value of the $z$ statistic.

• If $z \geqslant z^*$ then the observation falls into the "Reject $H_0$" region of the plot.

Table 34.2.: Table of critical values for the one sided hypothesis $H_A : \mu > \mu_0$.

| $\alpha$ (right tail area) | .10 | .05 | .02 | 0.01 | 0.005 | 0.0025 | 0.002 | 0.001 |
|---|---|---|---|---|---|---|---|---|
| $z^*$ (critical value) | 1.2816 | 1.6449 | 2.0537 | 2.3263 | 2.5758 | 2.807 | 2.8782 | 3.0902 |
| Random Chance 1 in every | 10 | 20 | 50 | 100 | 200 | 400 | 500 | 1000 |

When this occurs, the $p$ value will also satisfy $p \leqslant \alpha$.

Because the observation is in the far tail of the curve, it will only occur rarely by random chance when $H_0$ is true. Therefore we classify this result as significant. **We reject $H_0$ and accept $H_A$.**

- If $z < z*$ then the observation falls into the "Do Not Reject $H_0$" region of the plot.

When this occurs, the $p$ value will also satisfy $p > \alpha$.

Since the observation does not fall into the tail, it is expected to occur by random chance fairly often by random chance. Therefore we do not think that it provides evidence against $H_0$. **We do not reject $H_0$.**

**One Sided Alternate Hypothesis $H_A : \mu < \mu_0$.** The one-sided alternative hypothesis $H_A : \mu < \mu_0$ is illustrated in fig. 34.2. The curve shows a standard normal curve assuming that $H_0$ is true. The shaded region has area $\alpha$, based upon a confidence level $C$, and the area gives the probability

$$\alpha = P(z \leqslant z^*)$$

Changing the size of the tail region is equivalent to changing the value of $\alpha$. Smaller tail regions are more convincing, but how small is sufficient is purely subjective. Table 34.3 summarizes the $\alpha$ values, critical values, and how often such an event is likely to occur by random chance, for this type of alternative hypothesis.

Once the $z$ statistic is calculated,

- If $z \leqslant z^*$ then the observation falls into the "Reject $H_0$" region of the plot.

FAQ 34. How Small is Small?

Figure 34.2.: Accept and reject region for $H_A\colon \mu < \mu_0$. The null hypothesis is rejected if the normalized observation falls into the shaded area.

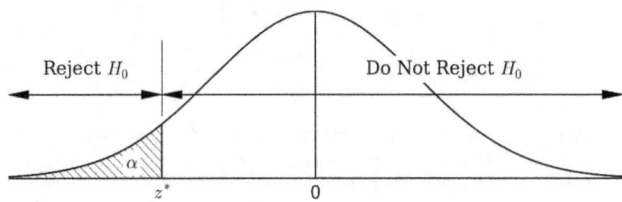

Table 34.3.: Table of critical values for the one sided hypothesis $H_A : \mu < \mu_0$.

| $\alpha$ (left tail area) | .10 | .05 | .02 | 0.01 | 0.005 | 0.0025 | 0.002 | 0.001 |
|---|---|---|---|---|---|---|---|---|
| $z^*$ (critical value) | -1.2816 | -1.6449 | -2.0537 | -2.3263 | -2.5758 | -2.807 | -2.8782 | -3.0902 |
| Random Chance 1 in every | 10 | 20 | 50 | 100 | 200 | 400 | 500 | 1000 |

When this occurs, the $p$ value will also satisfy $p \leqslant \alpha$.

Because the observation is in the far tail of the curve, it will only occur rarely by random chance when $H_0$ is true. Therefore we classify this result as significant. **We reject $H_0$ and accept $H_A$.**

- If $z > z*$ then the observation falls into the "Do Not Reject $H_0$" region of the plot.

When this occurs, the $p$ value will also satisfy $p > \alpha$.

Since the observation does not fall into the tail, it is expected to occur by random chance fairly often by random chance. Therefore we do not think that it provides evidence against $H_0$. **We do not reject $H_0$.**

**Two Sided Alternate Hypothesis** $H_A\colon \mu \neq \mu_0$. Fig, 34.3 illustrates the two-sided alternative hypothesis $H_A\colon \mu \neq \mu_0$ tested against the null hypothesis $H_0\colon \mu = \mu_0$ for a confidence level $C = 1 - \alpha$. The curve shows the standard normal curve assuming that $H_0$ is true, and the critical value $z^*$ is obtained from $\alpha$. The shaded region has area $\alpha$. Each of the two tails gives the probability

$$\frac{\alpha}{2} = P(z \geqslant |z^*|)$$

Figure 34.3.: Accept and reject region for the two sided alternative hypothesis $H_A$ : : $\mu \neq \mu_0$. The null hypothesis is rejected if the normalized observation falls into either shaded tail region.

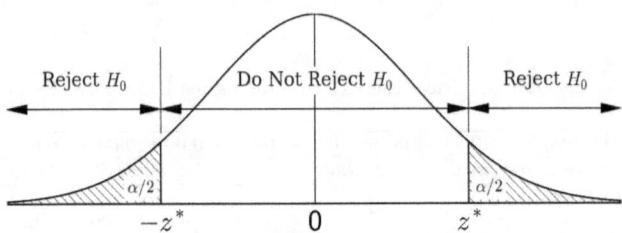

Changing the size of the tail region is equivalent to changing the value of $\alpha$. While smaller values are more convincing, the definition of sufficiently small is purely subjective. Table 34.4 summarizes the $\alpha$ values, critical values, and how often such an event is likely to occur by random chance, for this type of alternative hypothesis. Note that for a two sided hypothesis, the critical value $z^*$ is larger than the corresponding critical value for the same value of $\alpha$ for a one-sided hypothesis. This is because we must include two tail areas to obtain the same area, so we need to move further away from the mean to do this.

Table 34.4.: Table of critical values for the two sided hypothesis $H_A : \mu \neq \mu_0$.

| $\alpha$ (left tail area) | .10 | .05 | .02 | 0.01 | 0.005 | 0.0025 | 0.002 | 0.001 |
|---|---|---|---|---|---|---|---|---|
| $z^*$ (critical value) | 1.6449 | 1.96 | 2.3263 | 2.5758 | 2.807 | 3.0233 | 3.0902 | 3.2905 |
| Random Chance 1 in every | 10 | 20 | 50 | 100 | 200 | 400 | 500 | 1000 |

Once the $z$ statistic has been calculated from the data,

- If $z \leqslant -z^*$ or $z \geqslant z^*$ then the observation falls into one of the tail areas that are labeled as the "Reject $H_0$" regions of the plot.

FAQ 34.  How Small is Small?

When this occurs, the $p$ value will also satisfy $p \leqslant \alpha$.

Because the observation is in one of the far tail of the curve, it will only occur rarely by random chance when $H_0$ is true. Therefore we classify this result as significant. **We reject $H_0$ and accept $H_A$.**

- If $|z| < z*$, then the observation falls into the "Do Not Reject $H_0$" region of the plot.

When this occurs, the $p$ value will also satisfy $p > \alpha$.

Since the observation does not fall into *either* tail, it is expected to occur by random chance fairly often by random chance. Therefore we do not think that it provides evidence against $H_0$. **We do not reject $H_0$.**

# FAQ 35. How Do I Perform a Hypothesis Test on the Mean?

Here we will discuss the following question: how can we question the validity of the null hypothesis against an alternative?

We will only consider one format for the null hypothesis:

$$H_0 : \mu = \mu_0$$

Here $\mu$ is the population mean, and $\mu_0$ is a fixed constant. Thus **the null hypothesis will always state that the population mean is equal to some fixed value.**

We will consider three possible alternate hypotheses (ch. 32).

1. The one sided hypothesis: $H_A : \mu < \mu_0$.
2. The one sided hypothesis: $H_A : \mu > \mu_0$.
3. The two sided hypothesis: $H_A : \mu \neq \mu_0$.

Before you collect data you must state the significance level of your test. You do this by naming of value of $\alpha$ (ch. 34). Any observations that are collected that are deemed rare, in the sense that they have a probability $p \leqslant \alpha$, will be considered significant. Any other observations will not be considered significant, and will not be used to draw conclusions about the hypotheses. To review the details on the calculation of $p$ see chapter 33. For more on the distinction between $p$ and $\alpha$ see chapter 34.

Figure 35.1.: Interpretation of a $p$ value for the three different results, with the alternate hypothesis $H_A : \mu > \mu_0$. In the top frame, $z$ is small and the probability of falling in the shaded area is high. This is very likely to occur by random chance, and this does not provide any evidence against $H_0$. In the bottom frame, $z$ is large and the statistic falls very far from the mean. This statistic represents an observation that is unlikely to have occurred by random chance. The probability of falling in the shaded area is very small. This provides very strong evidence against $H_0$, and supports the alternative hypothesis. The frame in the middle is somewhere in-between.

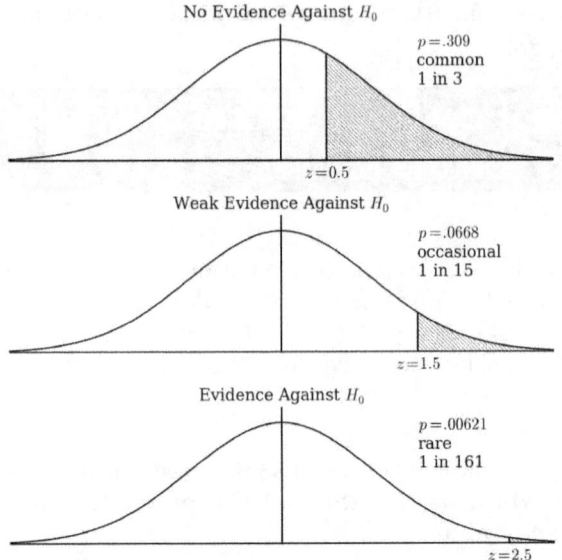

If we determine that the observed value of $\bar{x}$ is close enough to $\mu_0$ that it is fairly likely to occur by random chance, then our observation doesn't tell us anything.

However, if we determine that the observed value of $\bar{x}$ is extremely unlikely to occur when $H_0$ is true, then we say that it is a statistically significant (rare) event (as discussed in chapter 34).

In general, small $p$ values are evidence against $H_0$ and may be used to reject the null hypothesis and support the alternative hypothesis. They are said to be statistically significant evidence (at level $\alpha$) against $H_0$ when $p \leqslant \alpha$.

Table 35.1.: Interpretation of $z$-statistic and $p$-value.

| $z$ | $p$ | $H_0$ | $H_A$ |
|---|---|---|---|
| Large | Small $(p \leqslant \alpha)$ | Evidence Against Reject $H_0$ | Evidence For Accept $H_A$ |
| Small | Large $(p > \alpha)$ | No Conclusion | Not Supported |

The following steps summarize the procedure for preforming a hypothesis test on mean.

1. **State your question** in words.

   **Example 35.1.** You are told that the height of all American men is normally distributed with mean six feet and standard deviation 4 inches (0.33 feet). You suspect that American men really are not this tall so you will ask the following question: Is the average American shorter than 6 feet?

2. **Identify** the appropriate **parameter** (e.g., the population mean, $\mu$) and the corresponding statistic.

   (**Continuation of example 35.1.**) We will do a hypothesis test for the mean height, $\mu$, of all American men. The corresponding statistic will be a sample mean $\bar{x}$ of heights.

3. State the **null** ($H_0$) and **alternative** ($H_A$) **hypotheses.**

   (**Continuation of example 35.1.**) Since you suspect that men are really shorter than 6 feet tall, you formulate the following hypotheses:

   $$H_0: \mu = 6 \qquad \text{(American men are 6 feet tall)}$$
   $$H_A: \mu < 6 \qquad \text{(American men are shorter than 6 feet tall)}$$

4. Define what it means for your results to be **statistically significant.** In other words (see chapter 34), give a number $\alpha$ such that

   ▸ If you assume $H_0$ is true; and

- ▸ You calculate that $p < \alpha$; then
- ▸ You are willing to reject $H_0$

The second bullet point means that the observation or data is extremely unlikely to occur by random chance if $H_0$ is true.

(**Continuation of example 35.1.**) You decide that an event that occurs with one chance in 50 is unlikely to occur by random chance alone. Since $1/50 = 0.02$, you use $\alpha = 0.02$. This means that you will reject $H_0$ only if $p < 0.02$.

5. Obtain a random sample of size $n$, find its average value $\bar{x}$, and calculate the one sample **z statistic**

$$z = \frac{\bar{x} - \mu_0}{\sigma/\sqrt{n}}$$

where your null hypothesis has the form $H_0 : \mu = \mu_0$.

(**Continuation of example 35.1.**) You take a random sample of 36 adult men, and measure their height. You find that $\bar{x} = 70.5$ inches. Since $\mu_0 = 72$ inches (72 inches = 6 feet), $\sigma = 4$ inches (from item 1), and $n = 36$ (the sample size),

$$z = \frac{\bar{x} - \mu_0}{\sigma/\sqrt{n}} = \frac{70.5 - 72}{4/\sqrt{36}} = -2.25$$

6. **Find the p-value** (e.g. from A.1). The formulas for the p-value (as described in chapter 33) are shown in table 35.2.

Table 35.2.: Calculation of $p$ values for each of three different types of alternate hypothesis.

| Alternative Hypothesis | $p$ value |
|---|---|
| $H_A : \mu > \mu_0$ | $P(Z > z)$ |
| $H_A : \mu < \mu_0$ | $P(Z < z)$ |
| $H_A : \mu \neq \mu_0$ | $2P(Z > |z|)$ |

(**Continuation of example 35.1.**) Our hypothesis has the form $H_A : \mu < 6$, so $p = P(Z < -2.25)$. This is the tail area to the left of $z = -2.25$ under the curve of a standard normal distribution. From either table A.1 or software, this area is $p = .012$.

FAQ 35. Hypothesis Tests

7. **Compare p-value with $\alpha$.** If $p \leqslant \alpha$, reject $H_0$. If $p > \alpha$, do not reject $H_0$.

(**Continuation of example 35.1.**) We defined $\alpha = 0.02$ in step 4, and we determined that $p = 0.012$ in step 6. Therefore

$$p = 0.012 < 0.02 = \alpha$$

Since $p < \alpha$, our observation is statistically significant at the $\alpha = 0.02(2\%)$ level. Thus our observation is evidence against $H_0$ and supports the alternative hypothesis.

**Technology Example 35.2 (TI-84).** Repeat example 35.1 with a TI-84.

We use the **Z-Test** function for $\mu < \mu_0$.

1. Press [stat] then select [TESTS] 》 [Z-Test] [enter], or [TESTS] [1].

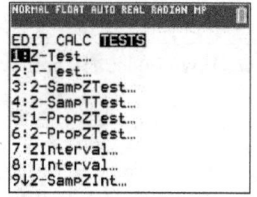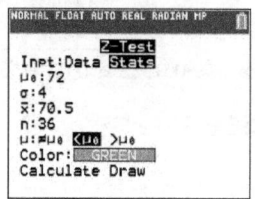

2. On the **Z-Test** menu set: **Inpt** to **Stats**; $\mu_0$ to 72; $\sigma$ to 4; $\bar{x}$ to 70.5; $n$ to 36; and select $\mu < \mu_0$.
3. Select **Calculate** and press [enter]. The $p$ value (0.0122244334) and $z$ score (-2.25) for the test appear on the screen.

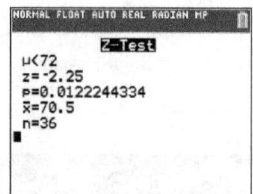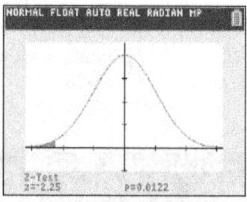

4. To plot the area corresponding by this test on a normal distribution, select **Draw** in step 3 instead of **Calculate**.

**Technology Example 35.3 (R).** Repeat example 35.1 in R.

Use **pnorm** with the **mean** and **sd** parameters, remembering to divide $\sigma$ by $\sqrt{n}$.

```
> pnorm(70.5, mean=72, sd=4/sqrt(36))
[1] 0.01222447
```

Since $p \approx 0.0122 < 0.02 = \alpha$ (given), we can reject $H_0$ and accept $H_A$.

**Technology Example 35.4 (Spreadsheet).** Repeat example 35.1 using a spreadsheet.

1. In any cell type the expression:

   =norm.dist(70.5, 72, 4/sqrt(36),1)

   followed by [enter]. The **norm.dist** function gives the area under a curve of a normal distribution; the arguments are:

   norm.dist(x, mean, sd, cum)

   where **cum** is set to 1 to indicate a cumulative distribution. We divide the standard deviation by $\sqrt{n}$ because the distribution represents the sample means. The answer (the p value) **0.012224473** appears in the cell after you hit [enter]:

| $f_x$ | =NORM.DIST(70.5, 72, 4/SQRT(36), 1) | | |
|---|---|---|---|
| C | D | E | F |
| 0.01222447 | | | |

2. Since $p \approx 0.0122 < 0.02 = \alpha$ (given), we reject $H_0$ and accept $H_A$.

# Exercises

Perform the following hypothesis tests. Are your results significant at (a) $\alpha = 5\%$ (b) $\alpha = 1\%$ (c) $\alpha = 0.1\%$?

1. $H_0 : \mu = 500$, $H_A : \mu > 500$, $\sigma = 100$, $\bar{x} = 510$, $n = 143$ (ans: z-value = 1.196; p-value = 0.116)
2. $H_0 : \mu = 23$, $H_A : \mu > 23$, $\sigma = 5.7$, $\bar{x} = 25$, $n = 74$ (ans: z-value = 3.018; p-value = 0.00254)

3. $H_0 : \mu = 500$, $H_A : \mu \neq 500$, $\sigma = 100$, $\bar{x} = 549$, $n = 17$ (ans: z-value = 2.020; p-value = 0.0433)
4. $H_0 : \mu = 10.7$, $H_A : \mu < 10.7$, $\sigma = 3.7$, $\bar{x} = 9.8$, $n = 75$ (ans: z-value = -2.107; p-value = 0.0176)
5. $H_0 : \mu = 64$, $H_A : \mu < 64$, $\sigma = 2.3$, $\bar{x} = 62.5$, $n = 20$ (ans: z-value = -2.917; p-value = 0.00177)
6. $H_0 : \mu = 480$, $\mu \neq 480$, $\sigma = 83$, $\bar{x} = 472$, $n = 1280$ (ans; z-value = -3.448; p-value = 0.000564)

# FAQ 36. What is the Difference Between Type I Error and Type II Error?

We can be wrong when we make a conclusion in a hypothesis test. There are two broad classes of error (table 36.1).

1. We might decide to reject $H_0$, even though $H_0$ is true. This is called **type 1 error** (or **type I error**). Type 1 error is also called a **False**

**Positive**, because when we reject $H_0$, we accept $H_A$, so we "test positive" for the alternative hypothesis. This is illustrated in fig. 36.1a. Suppose we observed some $\bar{x}$ value, labeled $\bar{x}^*$ in the figure. The p-value of our observation is shaded; if our p-value turns out to be significant, we will reject $H_0$.

Figure 36.1.: Illustration of the two types of error, showing distributions in the raw data domain. (a) Type I error (shaded). (b) Type II error. Both plots are drawn on the same horizontal axis for comparison.

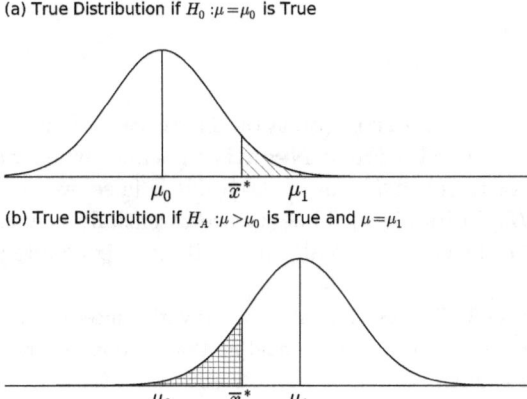

(a) True Distribution if $H_0 : \mu = \mu_0$ is True

$\mu_0$   $\bar{x}^*$   $\mu_1$

(b) True Distribution if $H_A : \mu > \mu_0$ is True and $\mu = \mu_1$

$\mu_0$   $\bar{x}^*$   $\mu_1$

**Example 36.1.** Sometimes a pregnancy test will give a positive result, e.g., will indicate that a woman is pregnant, even when she is not pregnant. The hypothesis is $H_0$:not pregnant; $H_A$ : pregnant. When the test shows that a woman is pregnant, even though she is not, we reject the null hypothesis and incorrect accept the alternate. This is a false positive, or type 1 error.

**Example 36.2.** The Prostate Specific Antigen (PSA) test is often used to test for prostate cancer in men. A level greater than 4.0 ng./ml is considered positive. In most men the level is close to zero. A possible hypothesis test might be $H_0 : \mu = 4.0$ vs. $H_A : \mu > 4.0$. If one tests positive ($\mu > 4$), then further testing is indicated. However, it turns that 75% of men who test postive actually have cancer. These are false positives. A typical value for $\mu_1$ might be $\mu_1 = 10$.

2. We might decide not to reject $H_0$, even though $H_0$ is false. This

---

Figure 36.2.: Comparison of $\alpha$, $\beta$, and power in the raw data domain.

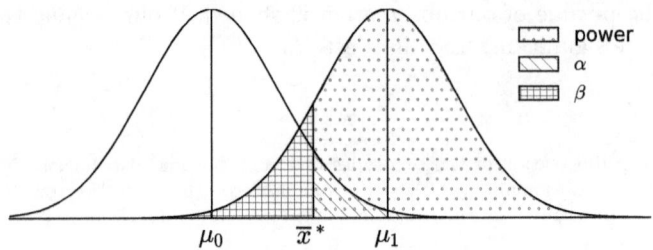

is called **type 2 error** (or **type II error**). Type 2 error is also sometimes called a **False Negative** because when we erroneously decide not to reject $H_0$, even though it is false, we do not choose to accept $H_A$ in its place. Thus we are essentially negating $H_A$ (even though, in fact, we are really just failing to conclude anything).

**Example 36.3.** Suppose that the patient with cancer in example 36.2 tests with $\bar{x} = 2.3$. The conclusion would be that he does not have cancer. This would be a false negative, or a type 2 error.

Table 36.1.: Classification of errors.

| Conclusion from sample | Truth About the Population | |
|---|---|---|
| | $H_0$ True | $H_0$ False |
| $H_A$ True | Type I Error<br>False Positive<br>Probability $\alpha$ | Correct Outcome<br>True Positive<br>Power |
| $H_A$ False | Correct Outcome<br>True Negative | Type II Error<br>False Negative<br>Probability $\beta$ |

The probability of a type 1 error is $\alpha$, the same $\alpha$ that we used to test for significance (chapter 34). Thus reduced the size of $\alpha$ (and at the same time reducing the size of the p-value that will be sufficient for a result to be considered significant), we also reduce the probability of a type 1 one error. The smaller the value of $\alpha$, the less likely we will have false positive results.

The probability of a type 2 error is labeled $\beta$. To actually calcuate $\beta$ we must have a value for the true mean $\mu_1$ when $H_A$ is true. $\beta$ is then the probability that we do not reject $H_A$ even though $H_A$ is false. The **power of the statistical test** is defined as $1 - \beta$.

The power and the vaues of $\alpha$ and $\beta$ are illustrated in fig. 36.2. During a hypothesis test, we first assume $H_0$ is true, i.e., the true mean is $\mu_0$. We calculate a critical value $z^*$. Corresponding to this critical value is a critical value of the mean, $\overline{x}^* = z^*\sigma + \mu_0$ (Note that we used the formula that assumes that $H_0$ is true). The power is the area under the curve of the distribution that assumes the mean $\mu_1$ is the correct mean, but to the right of $\overline{x}^*$. $\beta$ is the area under this same curve to the left of $\overline{x}^*$. By comparison, $\alpha$ is the area under the curve of the other distribution, the one that assumes $\mu_0$ is the correct mean, to the right of $\overline{x}^*$.

Figure 36.3.: Comparison of type 1 and type 2 areas. The shaded areas illustrate the trade-off between the minimization of $\alpha$ (top curve) and $\beta$ (bottom curve) for an alternate hypothesis of the form $H_A : \mu > \mu_0$. The horizontal axes are aligned in both curves for comparison. In general, we do not know the value of $\mu_A$ so the position of $\mu_A$ in the figure is shown for illustrative purposes; we only know that if $H_A$ is true then we expect that $\mu_A > z^*$. As $\mu_A$ moves to the right, the shaded area corresponding to $\beta$ gets smaller.

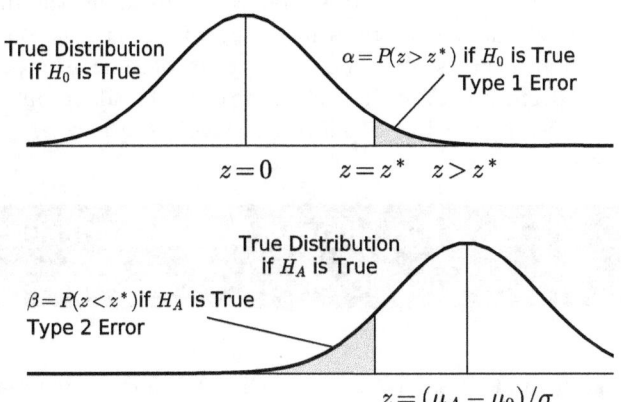

To design an effective research study, you need to know how large of a sample you need. Sample size depends on $\alpha$ and $\beta$. The calculation of $\beta$

and power are fairly complex and require that you make an assumption about the value of $\mu_A$. An ideal reasearch plan will minimize both $\alpha$ and $\beta$ while maximizing power. However this also requires a trade-off against sample size, since decreasing $\alpha$ and $\beta$ requires an increase in sample size. Larger samples are more expensive and this is an important consideration in research design. This subject is covered in more advanced texts and courses in statistics.

## FAQ 37. What is a t Distribution? What is the Difference Between a Normal and a t Distribution?

The **t-distributions** are really a family of different distributions rather than a single distribution. Like the normal distribution, the PDF (ch. 16) of each t distribution is bell shaped and symmetric with a single central peak. At first glance, a plot of single t-distribution looks almost exactly the same as a normal distribution. We distinguish between the different t-distributions by a parameter called the **degrees of freedom** or $df$ of the distribution. A t-distribution with $df$ degrees of freedom is sometimes written as $t_{df}$. The t-distributions are slightly wider than the normal distribution and do not approach zero as quickly for larger values of $x$ (fig. 37.1). When there are fewer degrees of freedom, the distribution is more spread out and there is more uncertainty in the data. When there are more degrees of freedom, the peak becomes sharper and the uncertainty decreases. As $df$ increases, the central peak becomes narrower and the shapes of the t-distributions resemble the normal distribution more closely.

## FAQ 38. How Do I Know Whether to Use a z-Statistic or a t-Statistic?

If we take a random sample of size $n$ from a normal population with known standard deviation $\sigma$, we should use the $z$ statistic. If we do not know $\sigma$, we can (in most cases) replace it with the sample standard deviation $s$, but in doing so, have to use a $t$ statistic with $n-1$ degrees of freedom. This is generally true even if the sample is not normal and $n$ is larger than around 40. If the sample is centrally peaked (mostly normal

Figure 37.1.: The $t$ distributions for two different degrees of freedom. The dotted line shows a standard normal distribution. As the number of degrees of freedom $(df)$ becomes very large, the $t$ distribution becomes very close to a normal distribution.

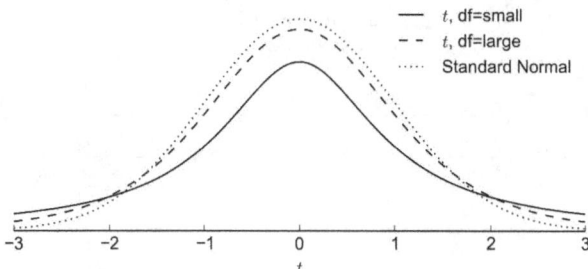

looking), this procedure usually works well for $n$ as small as 15. See figure 38.1. For smaller samples a sample correction needs to be applied, which is discussed in more advanced texts.

# FAQ 39. How Do I Find $t^*$?

The first part of solving any problem involving a t-distribution is figuring out the correct number of degrees of freedom. This is related to the size of the sample.

---

**Degrees of Freedom**

When using a t-distribution for a sample of size $n$, $df = n - 1$.

---

Confidence intervals and margins of error for $t$ statistics depend on **critical values** (fig 39.1). The critical value $t^*$ for a t-distribution is defined similarly to the critical value $z^*$ for a normal distribution (see page 94): the area bounded by the vertical lines $y = t^*$, $y = -t^*$, the $x$ axis, and the $t_{df}$ distribution is equal to $C$. To differentiate this critical value from the critical value for other degrees of freedom at the same confidence level we may sometimes express it as $t^*_{df}$.

Figure 38.1.: Decision tree for using the $t$ or $z$ statistic in inference of the mean. Here $\bar{x}$ is the sample mean; $s$ is the sample standard deviation; $n$ is the sample size; $\sigma$ is the population standard deviation; and $\mu_0$ is the population mean if the null hypothesis $H_0 : \mu = \mu_0$ is true.

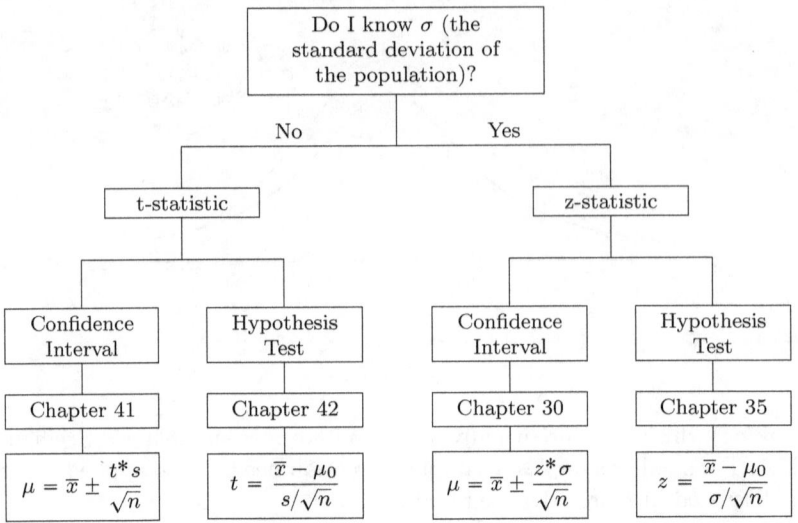

Critical values of the t-distribution are frequently tabulated (e.g., table A.2). The column headers list the confidence levels, and the row headers list the number of degrees of freedom (fig. 39.2). Each column then gives different critical values $t_{df}^*$ for the same confidence $C$; each row gives different critical values for different confidence at the same confidence. Here is how we use a table of this type:

1. The first column, on the left, gives the **possible degrees of freedom**. This is a list of integers that starts at 1, and increases to some large number, usually 100 or 1000. Except for the first 20 or so numbers, usually these are only in intervals of 10 or 20, up to 100, and then in intervals of 100 up to around 1000.
2. The numbers across the top row are different **confidence levels**.
3. The two rows at the bottom are p-values that are described in chapter 40. They are not used here.
4. The numbers in the interior of the table are the **critical values**. The critical value $t^*$ is found by locating the intersection of the row with the desired $df$ value and the column with desired confidence

Figure 39.1.: Critical values for the $t$ distribution.

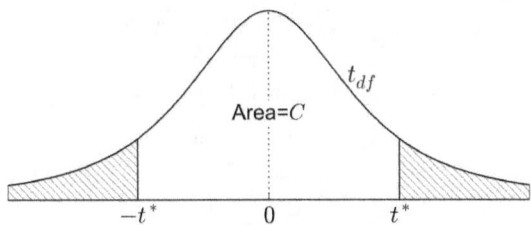

level (fig. 39.2). (If the value of $df$ you are looking for is not found, use the next smaller value in the table; this ensures that the margin of error will be sufficiently large.)

---

**Critical Value**

The **critical value** for a $t_{df}$ statistic with confidence level $C$ is the value $t^*$ such that the central area under the curve of the $t_{df}$ distribution, between $-t^*$ and $t^*$, is equal to $C$.

---

**Example 39.1.** Find the critical value $t^*$ when there are 12 degrees of freedom ($df = 12$) at a 98% confidence level.

Using table A.2 we look for at the number in the intersection of the row labeled 12 and the column labeled 98. This number is $t^* = 2.681$.

**Using a Calculator.** To use a calculator we need to invert a t-distribution. This can be a bit tricky, because we need to consider two tails. Since $C$ is the central area, the sum of the areas of **both tails** is $1 - C$. The area of **each tail** is therefore $(1 - C)/2$ and therefore the area to the left of the positive value $|t^*|$ is the sum of one tail and the center:

$$C + (1 - C)/2 = (1 + C)/2$$

---

Figure 39.2.: Layout of a typical table of the $t$ distribution (such as table A.2). The critical value is at the intersection of the row with the desired degrees of freedom ($df$) and the column with the desired confidence ($C$).

| df | | | C | | |
|---|---|---|---|---|---|
| | | Possible critical values | | | |
| 1 | ⋯ | nnn | ↓ | nnn | ⋮ |
| 2 | ⋯ | nnn | ↓ | nnn | ⋮ |
| | | ⋮ | ↓ | | |
| ⋮ | ⋯ | nnn | ↓ | nnn | ⋮ |
| df | → | → | $t^*$ | nnn | ⋮ |
| ⋮ | ⋯ | nnn | nnn | nnn | ⋮ |
| | | ⋮ | | | |
| 1000 | ⋯ | nnn | nnn | nnn | ⋮ |
| 1-Sided | ⋯ | aaa | aaa | aaa | ⋯ |
| 2-Sided | ⋯ | bbb | bbb | bbb | ⋯ |

---

## Area to the Left of Critical Value

The area to the left of the critical value $|t^*|$ is $(1 + C)/2$.

---

The consequence of this is that the critical value is the inverse of the t distribution for $(1+C)/2$.

---

## Critical Value (calculator)

The **critical value** for a $t_{df}$ statistic with confidence level $C$ is the inverse of the $t_{df}$ distribution at $x = (1 + C)/2$.

---

**Technology Example 39.2 (TI-84).** Repeat example 39.1 using a TI-84.

1. Press ⌷2nd⌷ ⌷distr⌷ and select ⌷invT⌷ ⌷enter⌷.
2. Set **area** to **(1+.98)/2** (this gives $(1 + C)/2$, which is required for the reason described above.)
3. Set **df** to **12**.

---

FAQ 39.   How Do I Find $t^*$?

4. Scroll to **Paste** then hit enter.
5. When you see **invT((1+.98)/2,12)** displayed, press enter. The answer **2.680997969** will be displayed.

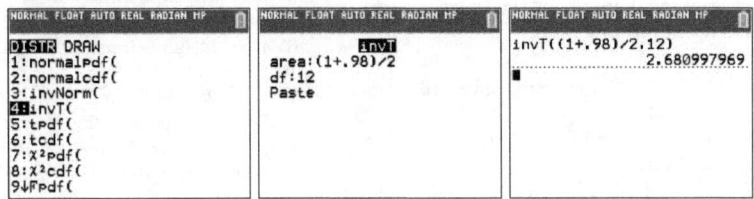

**Technology Example 39.3 (R).** Repeat example 39.1 using R.

We can using the **qt** function to invert the $t$ distribution. The format is **qt(x, df)**. The same cautions given for use with the calculator apply; we must set **x** to $(1 + C)/2$ to get the positive value $|t^*|$.

```
> qt((1+.98)/2,12)
[1] 2.680998
```

Alternatively, we could set **x** to $(1 - C)/2$ to get $-|t^*|$:

```
> qt((1-.98)/2,12)
[1] -2.680998
```

**Technology Example 39.4 (Spreadsheet).** Repeat example 39.1 using a spreadsheet.

Rather than inverting the $t$ distribution as a function of $C$ spreadsheets have a function that inverts it as a function of the probability $p = 1 - C$. The format is **t.inv.2t(p, df)**.

1. Type **=t.inv.2t(1-.98, 12)** into any cell and press enter.
2. The answer **2.680997993** appears in the same cell.

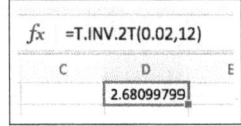

# Exercises

Find the degree of freedom $df$ and the critical numbers $t^*$ for the following confidence levels of $C$.

1. $n = 17$, 90% (ans: $df = 16$, $t^* = 1.746$)
2. $n = 5$, 80% (ans: $df = 4$, $t^* = 1.533$)
3. $n = 44$, 99% (ans: $df = 43$, not listed so use row 40 $t^* = 2.704$)
4. $n = 705$, 95% (ans: $df = 704$, not listed, so use row 100; $t^* = 1.984$)
5. $n = 73$, 95% (ans: $df = 72$, not listed, so use row 60; $t^* = 2.00$)
6. $n = 24$, 98% (ans: $df = 23$, $t^* = 2.5$)
7. $n = 28$, 98% (ans: $df = 27$, not listed, so use row 25; $t^* = 2.485$)
8. $n = 1432$, 96% (ans: $df = 1431 > 1000$ so use 1000, $t^* = 2.056$)

# FAQ 40. How Do I Find the p-Value for a t Distribution?

When you perform a hypothesis test using the one-sample t-statistic (ch. 42), the exact value of $t$ calculated is unlikely to occur in a table of the t-distribution such as table A.2. Instead, we use the following procedure (see fig. 40.1).

1. Locate the row corresponding to the number of degrees of freedom ($df$) in the problem. If the row is not present in the table, use the row with the largest number of degress of freedom less than $df$.

   **Example 40.1.** Find the one sided p-value corresponding to $t = 2.37$ with 9 degrees of freedom.
   Since $df = 9$, we use row 9 in the table. (continued)

2. Proceeding from left to right in the identified row, locate the first two values that bracket $t$.

   (**Continuation of example 40.1.**) As we proceed proceed in a rightwards direction along row 9, we see that 2.398 is the first value that exceeds $t$, and 2.262, just to its left, its the largest value that is less than $t = 2.37$.
   $$\boxed{df=9} ----> \boxed{2.262} < 2.37 < \boxed{2.398}$$

3. Following the columns containing the two bracketed values to the appropriate row in the bottom of the table.

   a) For a one-sided alternate hypothesis, proceed to the two values in the row labeled **one-sided p-value**.

(**Continuation of example 40.1.**) The bracketing values are 0.025 and 0.02. Therefore the one-sided p-value is bracketed by

$$0.02 < p < 0.025$$

   b) For a two-sided alternate hypothesis, proceed to the two values in the row labeled **two-sided p-value**.

4. The larger of the two bracketing values is the limiting p-value.

(**Continuation of example 40.1.**) We conclude that $p < 0.025$.

Figure 40.1.: Layout of a typical table of the $t$ distribution, such as table A.2. The normalized t-score with $df$ degrees of freedom will fall between two values $T_1$ and $T_1$ on line $df$. Follow the bounding values to the appropriate row on the bottom to get (upper and lower) bounding $p$ values.

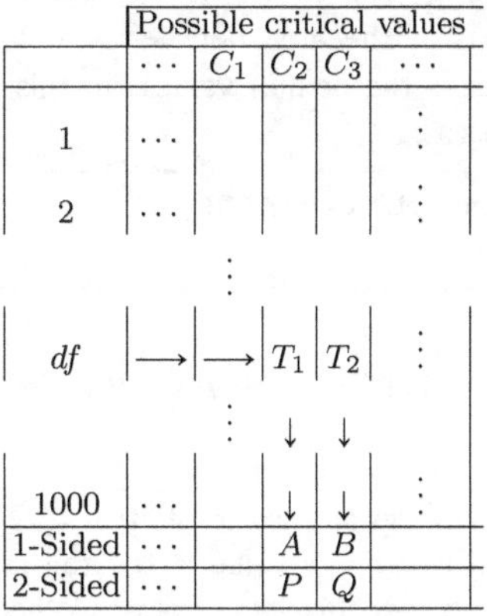

**Technology Example 40.2 (TI-84).** Repeat example 40.1 using a TI-84.

The one sided p value is the tail area. We will use **tcdf** to calculate this area. When $t$ is positive, it is the area under the curve to the right of $t$, i.e., between $t$ and $\infty$ (represented in the calculator by **1E99**). When $t$ is negative, it is the left tail area, or the area between $-\infty$ (represented as **-1E99**) and $t$. In this case, $t = 2.37$ is positive.

1. Open the distributions menu with [2nd] [var].
2. Select [tcdf] [enter] to select the cumulative distribution function for a t-distribution.

3. Set **lower** to **2.37**; **upper** to **1E99**; **df** to **9**.
4. Scroll to **Paste** and press enter.
5. When you see **tcdf(2.37, 1E99, 9)**, press enter.
6. The answer of **0.0209544812** will be displayed on the screen.

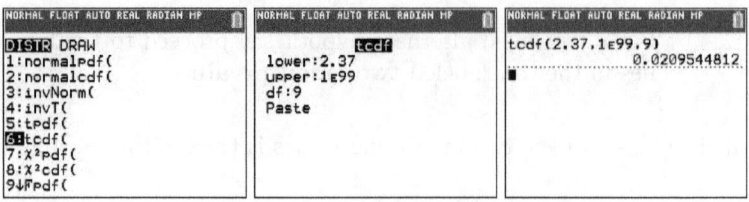

**Technology Example 40.3 (R).** Repeat example 40.1 using R.

The area under the curve of a t distribution is give by **pt**. When $t$ is positive, the p-value is the area under the curve to the right of $t$, i.e., between $t$ and $\infty$, obtained in R by setting **lower.tail=FALSE**. When $t$ is negative, the p value is the left tail area, or the area between $-\infty$ and $t$, obtained by setting **lower.tail=TRUE** (this is the default value). In this case, $t = 2.37$ is positive, so we set **lower.tail=TRUE**.

```
> pt(2.37,9,lower.tail=FALSE)
[1] 0.02095448
```

**Technology Example 40.4 (Spreadsheet).** Repeat example 40.1 using a spreadsheet.

The function **t.dist** gives the area to the left of $t$. The syntax is **t.dist(t, df, 1)**. When $t$ is positive, the $p$ value is the tail area to the right of $t$, i.e., between $t$ and $\infty$, so we subtract this value from 1. When $t$ is negative, the p value is the left tail area, or the area between $-\infty$ and $t$, so **t.dist** gives the p-value directly. In this case, $t = 2.37$ is positive, so we will have to subtract from 1.

1. In any spreadsheet cell, type **=1 - tdist(2.37, 9, 1)** enter.
2. The answer of **.02095448** will appear in the same cell.

# Exercises

Find the two numbers bracketing the given $t$ value in table A.2, and then bound the p-value.

1. $n = 10$, $t = 1.85$, one-sided (ans: df=9; $1.833 < t < 2.262$; $p < .05$)
2. $n = 17$, $t = 2.496$, one-sided (ans: df=16; $2.235 < t < 2.583$; $p < 0.020$)
3. $n = 42$, $t = 1.58$, one-sided (ans: df=41; not in table, so use df=40; $1.303 < t < 1.684$; $p < 0.1$)
4. $n = 23$, $t = 2.688$, two-sided (ans: df=22; $2.508 < t < 2.819$; $p < .02$)
5. $n = 7$, $t = 9.604$, one-sided (ans; df=6; $5.208 < t$; no upper limit to $t$ in table; $p < .001$)
6. $n = 1217$, $t = 1.73$, two-sided (ans: df=1216; largest value in table in 1000, so use 1000; $1.646 < t ; 1.962$; $p < .1$)

# FAQ 41. How Do I Find a Confidence Interval When I Don't Know the Population Standard Deviation?

You can still calculate a confidence interval on the mean of a normal population, even if you don't know the standard deviation of the population. You use the **t-distribution** instead of **z-distribution**.

Suppose you take a random sample of a population. Your population is centrally peaked but not necessarily normal. If you don't know the standard deviation of the population $\sigma$, you can still find a confidence interval on the mean of the population, so long as your population is not strongly skewed and your sample size is at least 40. If you sample is only slightly skewed then this procedure will usually work for samples as small as 15.

Here is procedure to find a level $C$ confidence interval:

1. Calculate the **sample mean** $\bar{x}$ (ch. 13).

> **Example 41.1.** Ramen noodles have become a staple of student diets around the world because they are inexpensive and easy to prepare. Here is a sample of Ramen Noodle prices, in dollars, in various supermarkets in southern California, per package,
>
> $0.19 $0.29 $0.45 $0.49 $0.23
> $0.13 $0.11 $0.17 $0.21 $0.29
> $0.22 $0.21 $0.23 $0.29 $0.34
>
> Find a 95% confidence interval on the mean cost per package of noodles.

We have $n = 15$ samples. The sample mean is

$$\bar{x} = \frac{1}{n}\sum x_i = 25.67 \text{ cents}$$

2. Calculate the **sample standard deviation** $s$ (ch. 15).

   (**Continuation of example 41.1.**)

   $$s = \sqrt{\frac{1}{n-1}\sum(x_i - \bar{x})^2} = 10.65 \text{ cents}$$

3. Calculate the **standard error of the sample mean**: sem $= \frac{s}{\sqrt{n}}$.

   (**Continuation of example 41.1.**)

   $$(\text{sem}) = \frac{s}{\sqrt{n}} = \frac{10.65}{\sqrt{15}} = 2.749$$

---

### Standard Error of the Sample Mean

The **standard error of the sample mean** is given by

$$(\text{sem}) = \frac{s}{\sqrt{n}}$$

where $n$ is the sample size and $s$ is the sample standard deviation.

---

4. Find the **critical value** $t^*$ in an t-distribution with $df = n - 1$ degrees of freedom corresponding to $C$. These are the values such that the central area between $t^*$ and $-t^*$ is equal to $C$ (ch. 39).

   (**Continuation of example 41.1.**) There are $n = 15$ observations, so there are $df = n - 1 = 14$ degrees of freedom. We find the critical value by looking in a t-table (such as table A.2). From row 14, column 95 (for the requested 95% confidence level), we find that $t^* = 2.145$.

5. Calculate the **confidence interval** (see box).

   (**Continuation of example 41.1.**) The confidence interval is

   $$\mu = \bar{x} \pm t^* \times (\text{sem}) = 25.67 \pm (2.145)(2.749) = 25.67 \pm 5.90$$

   or 19.77 to 31.59. Rounding to the nearest cent we get a confidence interval of 20 to 32 cents. Thus if the sample is repeated many times, we expect the

---

average value of a package of Ramen to fall between 20 cents and 32 cents 95% of the time.

> ### Confidence Interval on Mean
>
> **A level C confidence interval on the mean $\mu$ is**
>
> $$\mu = \bar{x} \pm t^* \times (\text{sem}) = \bar{x} \pm \frac{t^* s}{\sqrt{n}}$$
>
> where $z^*$ is the critical value, defined such that the central area under a t-distribution with $n-1$ degrees of freedom, between $z^*$ and $-z^*$, is $C$.

**Technology Example 41.2 (TI-84).** Repeat example 41.1 using a TI-84.

1. Enter the list of data items into a list.
   a) stat then select EDIT ⟩ Edit and enter.
   b) If there is already data in the **L1** column, you can erase it by navigating the cursor to the **L1** and pressing clear followed by enter.
   c) Enter each number followed by enter into the **L1** column.

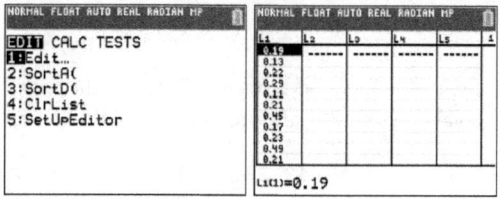

2. stat then select Tests ⟩ TInterval and enter.
   a) Set **Inpt** to **Data** and press enter.
   b) If necessary, set **List** to **L1** and press enter.
   c) If necessary, set **Freq** to **1** and press enter.
   d) If necessary, set **C-Level** to **0.95** and press enter.
   e) Navigate to **Calculate** and press enter. Then interval, mean, and standard deviation appear on the screen: **(0.1977, 0.3156)** (i.e., 19.77 to 31.56 cents); $0.25667 (25.667 cents); and $0.10648 (10.648 cents).

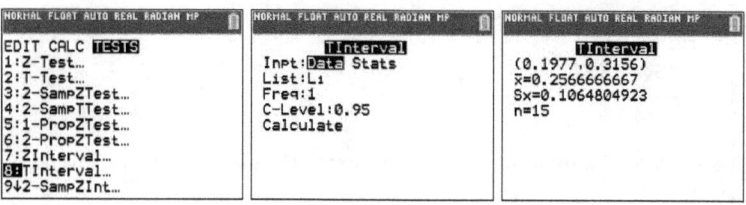

---

**Technology Example 41.3 (TI-84).** Repeat example 41.1 using a TI-84, assuming that you already know the values of $s_x = 10.65$ cents and $\bar{x} = 25.67$ cents.

1. stat then TESTS ⟩ Tinterval and enter.
2. Set **Inpt** to **Stats** and press enter.
3. Set $\bar{x}$ to **25.67**.
4. Set **Sx** to **10.65**.
5. Set **n** to **15** (the number of data items).
6. Set **C-Level** to **0.95**.
7. Navigate to **calculate** and press enter. The interval **(19.772, 31.568)** is displayed on your screen.

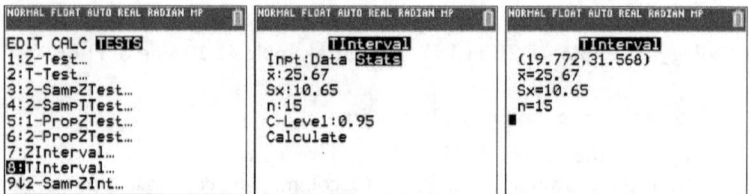

**Technology Example 41.4 (R).** Repeat example 41.1 using R.

We can do this with the function **t.test**. Entering the data in cents gives a 95% confidence interval of 19 to 33 (after rounding), as shown here.

```
> ramen=c(19,29,45,49,23,13,11,17,21,29,22,21,23,29,
  34)
> t.test(ramen,conf.level=0.95)

        One Sample t-test

data:  ramen
t = 9.3357, df = 14, p-value =2.174e-07
alternative hypothesis: true mean is not equal to 0
95 percent confidence interval:
 19.76997 31.56336
sample estimates:
mean of x
 25.66667
```

**Technology Example 41.5 (Spreadsheet).** Repeat example 41.1 using a spreadsheet.

---

The function **confidence.t** gives the margin of error using a t-distribution. To use it one must know the sample size, the standard deviation, and $\alpha = 1 - C$. The full syntax is **=confidence.t(alpha, s, n)** where **sd** is the sample standard deviation and **n** is the sample size.

| fx | =AVERAGE(C1:C15)+CONFIDENCE.T(0.05, STDEV(C1:C15), COUNT(C1:C15)) | | | | | | |
|---|---|---|---|---|---|---|---|
| C | D | E | F | G | H | I | |
| 19 | Lower: | 19.7699736 | | | | | |
| 29 | Upper: | 31.5633598 | | | | | |
| 45 | | | | | | | |
| 49 | | | | | | | |
| 23 | | | | | | | |
| 13 | | | | | | | |
| 11 | | | | | | | |
| 17 | | | | | | | |
| 21 | | | | | | | |

1. Enter the data into any column of the spreadsheet, say cells **C1:C15**.
2. The standard deviation can be found using **stdev(C1:C15)**.
3. The length can be found using **count(C1:C15)**.
4. The margin of error is
   **confidence.t(0.05, STDEV(C1:C15), COUNT(C1:C15))**
5. The mean is **average(C1:C15)**
6. The lower end of the confidence interval is
   **=average(C1:C15)**
   **-confidence.t(0.05, STDEV(C1:C15), COUNT(C1:C15))**
7. The upper end of the confidence interval is
   **=average(C1:C15)**
   **+confidence.t(0.05, STDEV(C1:C15), COUNT(C1:C15))**

# Exercises

Find the sample standard deviation, given the following sample sizes and standard error of the mean.

1. $n = 25$, (sem)$=.25$ (ans: $s = 25$)
2. $n = 7$, (sem)$=.4$ (ans: $s = 1.058$)
3. $n = 742$, (sem)$=3.5$ (ans: $s = 95.3$)
4. $n = 85$, (sem)$=7.2$ (ans: $s = 66.4$)

Find the standard error of the mean for the given sample standard deviations and sample sizes.

5. $s = 507$, $n = 23$ (ans: (sem)$=105.7$)
6. $s = 515$, $n = 840$ (ans: (sem)$=17.8$)
7. $s = 15$, $n = 35$ (ans: (sem)$=2.54$)
8. $s = 12$, $n = 5$ (ans: (sem)$=5.37$)

Find confidence intervals of the mean give the following sample data sets. give answers to one decimal point accuracy.

9. C=95% for:
   504, 619, 554, 647, 524, 538, 559
   (ans: 515.9 to 611.2)
10. C=98% for:
    38, 34, 37, 37, 34, 36,
    37, 34, 35, 36, 30, 34
    (ans: 36.6 to 37.7)
11. C=90% for :

72, 77, 59, 71, 48, 75, 76, 63, 61,
79, 68, 73, 70, 76, 73, 70, 74, 64,
81, 85, 77, 66, 51, 71, 67, 59, 63,
78, 63, 61
(ans: 66.4 to 71.7)

12. C=99.8% for:
27, 12, 34, 29, 23, 16, 28, 11, 20,
38, 34, 21, 28, 37, 34, 20, 30
(ans: 18.5 to 33.5)

# FAQ 42. How Do I Do a Hypothesis Test When I Don't Know Standard Deviation of the Population?

Suppose you take a random sample from some population. The population is not necessarily normal, and you don't necessarily know its standard deviation. If the distribution is relatively symmetric and centrally peaked, you can still do hypothesis test using the **one-sample t-statistic**. This procedure is generally valid for samples larger than around 40. If the distribution is not very heavily skewed, it can be used for samples as small as 15.

We form alternate hypotheses in the same manner as we did when we were sampling a normal population and knew the standard deviation (ch. 32, 35). As before, we will consider three possible alternate hypotheses: $H_A: \mu > \mu_0$; $H_A: \mu < \mu_0$, and $H_A: \mu \neq \mu_0$. The p-values for each type of alternate hypothesis correspond to the same tail as before, except this time it is the tail of an appropriate $t$-distribution. The procedure for calculating the p-value from a table of the t-statistic is given in ch. 40.

### One Sample t-Statistic

The **one-sample t-statistic** for testing against the null hypothesis $H_0: \mu = \mu_0$ is

$$t = \frac{\overline{x} - \mu_0}{s/\sqrt{n}}$$

where $s$ is the sample standard deviation and $n$ is the sample size. The t-statistic has n-1 degrees of freedom.

Here is the procedure for performing a hypothesis test.

1. State your **question** in words.

   **Example 42.1.** You are told that the typical (average) height of American men is six feet. You suspect that American men are really are somewhat shorter, so you ask the following question: Is the average American shorter than 6 feet?

2. Identify the appropriate **parameter** (e.g., the population mean, $\mu$) and the corresponding statistic.

   (**Continuation of example 42.1.**) The population is all American men; the corresponding parameter is their mean height $\mu$. Our statistic will be mean height $\bar{x}$ of a sample of American men.

3. State the null ($H_0$) and alternative ($H_A$) **hypotheses**.

   (**Continuation of example 42.1.**) Since you suspect that men are really shorter than 6 feet tall, you formulate the following hypotheses:

$$H_0 : \mu = 6 \qquad \text{(American men are 6 feet tall)}$$
$$H_A : \mu < 6 \qquad \text{(American men are shorter than 6 feet tall)}$$

---

### Guideline for Choosing $\alpha$

- Tests with smaller $\alpha$ will give results that you are more certain are correct (you are more confident in).
- If $\alpha$ is too small your test is likely to reject data that should have supported $H_A$.
- If $\alpha$ is too large your test is likely to accept data that really does not support $H_A$, but is, in fact, due to random chance.
- The most common value for $\alpha$ is 0.05 (5%).

---

4. Define what it means for your results to be **statistically significant**. In other words (see chapter 34), give a number $\alpha$ such that

   - If you assume $H_0$ is true; and
   - You calculate that $p < \alpha$; then
   - You are willing to reject $H_0$

   The second bullet point means that the observation or data is extremely unlikely to occur by random chance if $H_0$ is true.

   (**Continuation of example 42.1.**) You decide that an event that occurs with one chance in 50 is unlikely to occur by random chance alone. Since

---

$1/50 = 0.02$, you use $\alpha = 0.02$. This means that you will reject $H_0$ only if $p < 0.02$. In general, **the choice of what value to use for $\alpha$ is completely arbitrary and up to you.**

5. Take a **sample** of size $n$, find its **mean** $\overline{x}$ (ch. 13).

   **(Continuation of example 42.1.)** You take a random sample of adult men, and measure their height. Here are your measurements in inches:

   $$73.55,\ 70.23,\ 70.94,\ 67.44,\ 66.31,\ 68.73,\ 67.42,$$
   $$69.13,\ 73.10,\ 73.73,\ 68.42,\ 66.40,\ 68.02,\ 68.05$$

   You find that $\overline{x} = \dfrac{1}{n}\sum x_i = 69.39$ inches.

6. Calculate the **sample standard deviation** $s$ (ch. 15).

   **(Continuation of example 42.1.)**

   $$s = \sqrt{\frac{\sum(x_i - \overline{x})^2}{n-1}} = 2.55$$

7. Calculate the **one-sample t-statistic** $t = \dfrac{\overline{x} - \mu_0}{s/\sqrt{n}}$.

   **(Continuation of example 42.1.)**

   $$t = \frac{\overline{x} - \mu_0}{s/\sqrt{n}} = \frac{69.39 - 72}{2.55/\sqrt{14}} = -3.83$$

8. Find the **p-value** (table A.2; ch. 40). Use the fact that the t-statistic has $df = n - 1$ degrees of freedom.

   **(Continuation of example 42.1.)** Since there are $n = 14$ observations, the t-statistic has $df = 14 - 1 = 13$ degrees of freedom. Looking in table A.2, we see that $t = 3.83$ is bounded by 3.012 and 3.852 in line 13. Following these numbers down to the bottom of the the table gives $0.001 < p < 0.005$ (this is a one-sided hypothesis test). Using the larger of the bounding p-values gives a p-value of 0.005.

9. Compare the p-value with $\alpha$. If p-value $\leqslant \alpha$, reject $H_0$. If the p-value $> \alpha$, do not reject $H_0$.

   **(Continuation of example 42.1.)** Since $\alpha = 0.02$ and the p-value $= 0.005$, we conclude that the p-value $< \alpha$. Thus we have a statistically significant at the $\alpha = 0.02$ (2%) level. This is evidence against $H_0$ and supports the alternative hypothesis: with 98% confidence, we conclude that American men are shorter than six feet.

---

FAQ 42. Hypothesis Testing on the Mean (t-test)

**Technology Example 42.2 (TI-84).** Repeat example 42.1 using a TI-84.

1. Enter the data into list **L1**.
   a) Type stat followed by EDIT ⟩⟩ Edit and then enter.
   b) Navigate to the **L1** and type the data items in one at a time, following each one with the enter. If list **L1** is taken, you may either select another list, or clear the list. To clear the list, navigate to the label **L1** and press clear followed by enter.

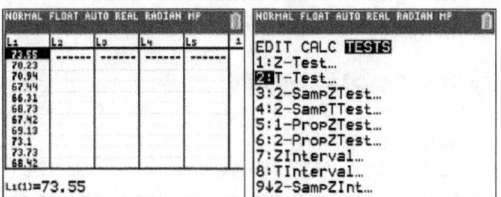

2. Once the data is entered in a list, go to the T-tests page. Hit stat followed by TEST ⟩⟩ T-Test and then enter.
3. On T-test data entry page,
   a) Select **Data** (not **Stats**) and press enter.
   b) Set $\mu_0$ to 72.
   c) Set **List** to **L1**, unless you changed lists. To select a different list, tpe 2nd 2 for **L2**, 2nd 3 for **L3**, etc., here.
   d) Set **Freq** to **1**
   e) In the row labeled $\mu$ select $< \mu_0$ to select the appropriate alternative hypothesis.
4. Scroll to **Calculate** and press enter.

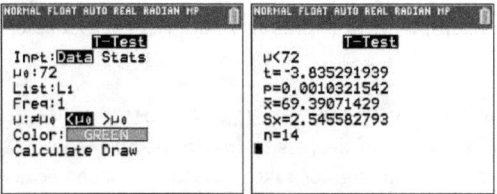

5. To make a plot return to the data entry page, and select the desired color. Then scroll to **Draw** and press enter.

**Technology Example 42.3 (R).** Repeat example 42.1 in R.

We can use the **t.test(data)** function, where **data** is a vector containing the data. The full syntax we need to use is

```
t.test( data, alternative="value", conf.level=value,
                   mu=value)
```

where **alternative** is set to a value of **"less"** (representing $H_A : \mu < \mu_0$), **"greater"** (representing $H_A : \mu > \mu_0$), or **"two-sided"** (representing $H_A : \mu \neq \mu_0$); **conf.level** is set to the confidence value $C$ (between 0 and 1); and **mu** is set equal to the value of $\mu_0$ in the hypothesis.

```
> heights=c(73.55, 70.23, 70.94, 67.44, 66.31,
    68.73, 67.42, 69.13, 73.10, 73.73, 68.42,
    66.40, 68.02, 68.05 )
> t.test(heights, alternative="less",
    conf.level=0.98, mu=72.0)


        One Sample t-test

data:  heights
t = -3.8353, df = 13, p-value = 0.001032
alternative hypothesis: true mean is less than 72
98 percent confidence interval:
      -Inf 70.94297
sample estimates:
mean of x
 69.39071
```

The R code tells us that the alternative hypothesis has a 98% confidence interval of $[-\infty, 70.94297]$; since $\bar{x} = 69.39071$ falls within this interval, we accept the alternative hypothesis and reject the null hypothesis. We can make the same conclusion by observing that $p = .001032 < .02 = 1 - C = \alpha$.

**Technology Example 42.4 (Spreadsheet).** Repeat example 42.1 using a spreadsheet.

We can do this using the function **tdist(t, df, 1)**, where the arguments are the t-value, the number of degrees of freedom, and the number of tails. When $t$ is negative, we put a minus sign or take its absolute value because of implementation restrictions (negative values not allowed).

1. Enter the data into a column. Say, for example, the data is in cells **C1:C4**.
2. Enter the following formula into any cell:

```
=TDIST(-(AVERAGE(C1:C14)-72)/
 (STDEV(C1:C14)/SQRT(COUNT(C1:C14))),
 COUNT(C1:C14)-1,
 1)
```

Here the first two lines calculate the t-value as $(\bar{x} - \mu_0)/(s/\sqrt{n})$, using $\mu_0 = 72$. The second line calculates the degrees of freedom as $n - 1$.

3. Pressing ⌊enter⌋ the p-value of 0.001 appears in the same cell.
4. Since $\alpha = 0.02$ and $p < \alpha$, we accept the alternative hypothesis and reject the null.

| | Edit Links | ∧ | | | ∨ | | Columns | | | | |
|---|---|---|---|---|---|---|---|---|---|---|---|

*fx* =TDIST(-(AVERAGE(C1:C14)-72)/(STDEV(C1:C14)/SQRT(COUNT(C1:C14))), COUNT(C1:C14)-1,1)

| C | D | E | F | G | H | I | J | K |
|---|---|---|---|---|---|---|---|---|
| 73.55 | 0.001032154 | | | | | | | |
| 70.23 | | | | | | | | |
| 70.94 | | | | | | | | |
| 67.44 | | | | | | | | |
| 66.31 | | | | | | | | |
| 68.73 | | | | | | | | |
| 67.42 | | | | | | | | |
| 69.13 | | | | | | | | |
| 73.1 | | | | | | | | |

# Exercises

1. Suppose that you get back a pyschology exam in a mega-class with several hundred students. The professor said that anyone who scored at or above the mean will get a B, but did not tell you what the mean is. You heard that last year the mean was 80. Your test score was 78. You ask a number of students their scores and based on the results test the following hypothesis: $H_0 : \mu = 78$ vs $H_A : \mu < 80$. Here are their scores: 76.0, 69.0, 69.0, 69.0, 77.0, 84.0, 87.0
   (a) Find the degrees of freedom and appropriate t-statistic;
   (b) bound the p-value for the test;
   (c) what would you conclude
   (ans:$\bar{x} = 75.857$; $s = 7.448$; df=6; t $= -1.472$; $.7 < p < .8$; there is not enough evidence to conclude that the mean is less than 80 this year.)

2. Repeat the previous exercise with the following data set:
   70, 81, 78, 71, 76, 79, 74, 74, 88, 82, 84, 81, 80, 86, 73, 73, 77, 78, 78, 75, 68, 77, 84, 82, 82, 67, 75, 78, 84, 82, 82, 77, 74, 72, 82
   (ans: $\bar{x} = 77.829$; $s = 5.096$; t $= -2.521$; df $= 34$)

# FAQ 43. Should I Do Matched Pairs or a Difference of Means?

A **matched pairs** study (ch. 45, 44) can only be performed when you have data for the **same sample under different conditions**.

**Example 43.1.** To determine if the math SAT or the verbal SAT is harder, you select a random sample of 154 college students who have taken both parts of the exam. You compute the difference of each students scores, verbal − math. A positive score indicates that math is harder, while a negative score indicates that verbal is harder. This is a matched pair study because there is a single sample from a single population (students who took both tests) and we have two results for each individual (fig. 43.1).

Figure 43.1.: Illustration of matched pairs comparison discussed in example 43.1.

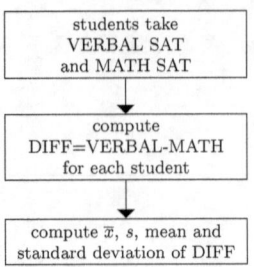

A **difference of means** study (ch. 46, 47, 48) should be performed when you are **comparing samples from two different populations**. The samples should be chosen independently of one another, i.e., the probability of being in one sample should not depend on the probability of being in the other sample.

**Example 43.2.** To determine whether the Graduate Record Exam (GRE) or the Law School Admission Test (LSAT) is more difficult, you collect two samples. One is a sample of 25 students who took the math GRE. The other is a sample of the 15 students who took the LSAT. You calculate the sample means compare the difference. This is a difference of means study because the data comes from two independent populations: students taking the GRE and students taking the LSAT. While it is possible that some students can be in both groups, the existence of a

Figure 43.2.: Illsutration of difference of means comparison described in example 43.2.

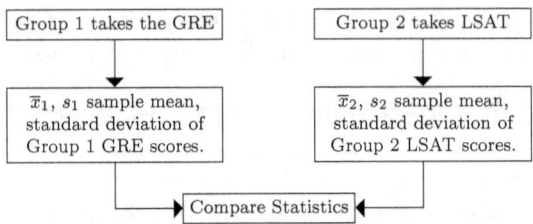

student in one group does not increase or decrease his or her likelihood of being in the other group, so the two populations are still independent. Also, we notice that the two samples have different sizes, which is not possible in a matched pair study (fig. 43.2).

Either a confidence interval or a hypothesis test can be performed on either type of study.

# FAQ 44. How Do I Find a Confidence Interval Using Matched Pairs Data?

Suppose we take a single random sample of individuals that has been subjected to two different conditions. We have the results of some variable for each of these individuals after each condition. An examination of the differences of the results is called a matched pair study.

## Confidence Interval on Matched Pairs

A level C confidence interval on the mean of the differences in a matched pair study is

$$\mu = \bar{x} \pm t^* \times (\textbf{sem})$$

where $\bar{x}$ is the mean of the sample difference; $(\text{sem}) = s/\sqrt{n}$ is the standard error of the mean; $s$ is the standard deviation of the sample differences; $n$ is the sample size; and $t^*$ is the critical number in a t-distribution with $n-1$ degrees of freedom, defined so that the central area under a $t_{n-1}$ distribution between $-t^*$ and $t^*$ is $C$.

**Example 44.1.** Twenty seven athletes begin a ten week training course in speed swimming. You measure how long it takes each swimmer to complete a 200 meter swim on the first day of the class, and then repeat the measurements at the end of the tenth week. You have two measurements for each swimmer.

We look at hypothesis testing in matched pair data in chapter 45. Here we wll look at the computation of a confidence interval on the difference.

1. Suppose your data is labeled as pairs$(u, v)$, where $u$ is the first mea-

surement or observation, and $v$ is the second. Calculate the differences $x_1 = v_1 - u_1$, $x_2 = v_2 - u_2$, $x_3 = v_3 - u_3$, ..., $x_n = v_n - u_n$.

(**Continuation of example 44.1.**)
Here are the swim times in seconds and improvements:

| first week | tenth week | change $v - u$ | first week | tenth week | change $v - u$ | first week | tenth week | change $v - u$ |
|---|---|---|---|---|---|---|---|---|
| 123 | 139 | 16 | 117 | 141 | 24 | 144 | 130 | -14 |
| 138 | 118 | -20 | 136 | 113 | -23 | 149 | 107 | -42 |
| 137 | 128 | -9 | 129 | 134 | 5 | 132 | 120 | -12 |
| 135 | 111 | -24 | 136 | 128 | -8 | 142 | 127 | -15 |
| 133 | 116 | -17 | 134 | 126 | -8 | 133 | 128 | -5 |
| 107 | 114 | 7 | 133 | 110 | -23 | 154 | 117 | -37 |
| 124 | 121 | -3 | 122 | 117 | -5 | 142 | 131 | -11 |
| 146 | 110 | -36 | 141 | 124 | -17 | 134 | 141 | 7 |
| 140 | 137 | -3 | 140 | 128 | -12 | 134 | 131 | -3 |

2. Find the **mean of the differences** $\bar{x}$ (ch. 13) and the **standard deviation of the differences** $s$ (ch. 15).

(**Continuation of example 44.1.**) Using the table of differences above, we find the $\bar{x} = -10.67$ and $s = 14.96$.

3. Find the **standard error of the mean** (sem) $= \dfrac{s}{\sqrt{n}}$.

(**Continuation of example 44.1.**) (sem) $= \dfrac{s}{\sqrt{n}} = \dfrac{14.96}{\sqrt{27}} = 2.88$

4. Find the **critical value** $t^*$ for the desired confidence (see ch. 39).

(**Continuation of example 44.1.**) Since $n = 27$ there are $n - 1 = 27 - 1 = 26$ degrees of freedom. Table A.2 does not have any lines between 25 and 30 so we use line 25. This gives $t^* = 2.06$.

5. Find the **margin of error** (me) $= t^* \times$ (sem).

(**Continuation of example 44.1.**) Using $t^* = 2.06$ and (sem) $= 2.88$ gives (me) $= (2.06) \times 2.88 = 5.93$

6. The **confidence interval** is $\bar{x} \pm$ (me).

(**Continuation of example 44.1.**) Using $\bar{x} = -10.67$ and (me) $= 5.93$ gives a 95% of confidence interval $-10.97 \pm 5.93$ seconds, or $-16.90$ to $-4.14$ seconds. Since the numbers are negative, the swimmers completed the laps in less time, so they swam faster. This represents an improvement.

We concluded that the average improvement was between 4.14 and 16.90 seconds with 95% confidence.

**Technology Example 44.2 (TI-84).** Repeat example 44.1 using a TI-84.

1. Load the data for the first week and the tenth week into into two lists, e.g., **L1** and **L2**, e.g., by typing or using the data cable to load the data from a PC.
2. Use list **L3** to calculate the difference.
   a) Open the list menu by hitting [stat] [Edit] [enter].
   b) Navigate the cursor to the **L3** header. To clear out any data that is currently in **L3**, press [clear] [enter].
   c) Type [L2] [-] [L1] [enter]. (To get [L2], use [2nd]+[2], etc.) This will calculate the difference in the entire list.

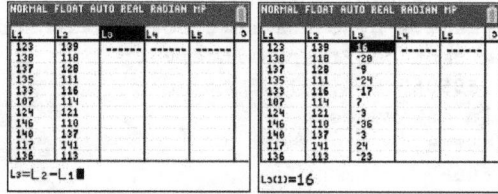

3. Select [stat] ⟩ TESTS ⟩ T-Interval [enter] to bring up the t-interval menu.
   a) For **Inpt:** select **Data**.
   b) Set **List:** to L3 (for the differences)
   c) Leave **Freq:** set to **1**
   d) Set the **C-Level** to **0.95** for 95% confidence.
   e) Scroll to **Calculate** and click **enter**

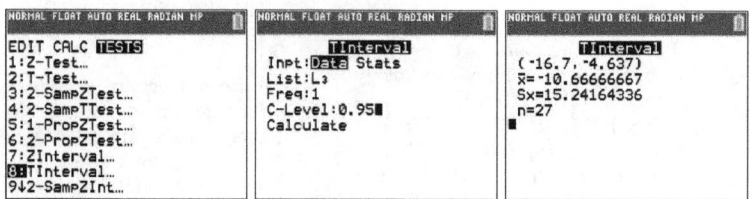

4. The confidence interval is displayed as −16.7 to −4.6 seconds.

**Technology Example 44.3 (R).** Repeat example 44.1 using R.

Put the two data sets into arrays and calculate the difference.

```
>u=c(123,138,137,135,133,107,124,146,140,117,
   136,129,136,134,133,122,141,140,144,149,132,
   142,133,154,142,134,134 )
```

```
>v=c(139,118,128,111,116,114,121,110,137,141,
   113,134,128,126,110,117,124,128,130,107,120,
   127,128,117,131,141,131)
> w=v-u
```

The function **t.test** is used to perform both hypothesis tests and calculate confidence intervals. The code **t.test(w)** will find a 95% confidence interval.

```
> t.test(w,conf.level=.95)

         One Sample t-test

data:  w
t = -3.6365, df = 26, p-value = 0.001197
alternative hypothesis: true mean is not equal to 0
95 percent confidence interval:
 -16.696060   -4.637273
sample estimates:
mean of x
-10.66667
```

This tells us that the 95% confidence interval on the differences is $-16.696060$ to $-4.637273$. This test also gives us the $p$-value for a hypothesis test of the two sided hypothesis $H_A : \mu \neq 0$, telling us that it has a $p$-value of 0.001197.

# Exercises

1. The following pairs give results of exams in a freshman psychology course as (Test1, Test2).
   (67, 85), (71, 81), (70, 62), (78, 79), (74, 71), (53, 78), (80, 73), (77, 67), (73, 62), (71, 65), (73, 76), (70, 64) Calculate 95% and 98% confidence intervals on the mean change in score.

2. Repeat the previous problem using the following data:
   (64, 67), (74, 81), (74, 76), (70, 74), (72, 72), (66, 81), (73, 81), (65, 70), (68, 69), (73, 72), (69, 79), (71, 80), (70, 74), (68, 68), (67, 83)

3. Fifteen men are sampled from a large group who are testing a new diet. Here are their weights in pounds (before, after):
   (204, 202), (189, 186), (181, 179), (210, 211), (195, 195), (203, 204), (220, 221), (224, 223), (203, 197), (211, 208), (207, 204), (215, 215), (189, 187), (199, 201), (219, 216)
   Find a 95% confidence interval on the weight change.

# FAQ 45. How Do I Perform a Hypothesis Test on Matched Pairs Data?

For a matched pairs calculation your sample data must be collected under two different conditions *for each individual in the sample.* You are not comparing test groups, but *the responses to different conditions of the same individuals.* For example, you might be comparing standardized test scores in English and math in the same individuals to see if students who do poorly in math also do poorly in English. A related technique is to compare the samples means, that is, to compare the average grades of math students (who may or may not study English) with the average grades of English students; (who may or not study math). In a matched pair study we would have to limit our population to only those students who studied both math and English.

The matched pair process is illustrated in fig. 45.1. Here we have defined $d_i$ as the difference between two observations of the same individual under different conditions. We call these observations $x_{i,1}$ and $x_{i,2}$. The mean of all the differences $d_1, d_2, \ldots, d_n$ is called $\overline{d_n}$. This is the sample statistic for a matched pair trial. It has a $t$ distribution with $n - 1$ degrees of freedom.

The number of degrees of freedom for the test is determined by the formula

$$df = n - 1.$$

---

### Matched Pair Statistic

To test the one sample hypothesis $H_0 : \mu = 0$, where $\mu$ is the average of the change in value in of the variables $x_{i,1} - x_{i,2}$ in the population, use a the $t_{df}$ statistic

$$t_{df} = \frac{\overline{x} - 0}{(\text{sem})} = \frac{\overline{x}}{s/\sqrt{n}}$$

on a random sample. Here $x_{i,1}$ and $x_{i,j}$ are the values of $x_i$ under the two different treatments or conditions; $\overline{x}$ is the sample average of $x_{i,1} - x_{i,2}$; and $n$ is the sample size.

---

Figure 45.1.: Concept of a matched pair experiment or observation. The same sample is observed under different conditions (ducks in the rain or ducks in the snow) or subjected to different treatments (aspirin or placebo).

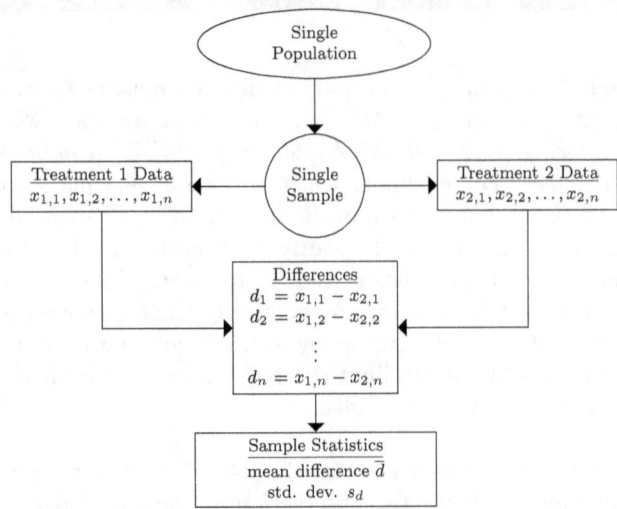

Here is the procedure for doing a matched pair calculation.

1. Call the two conditions $u$ and $v$.

   **Example 45.1.** Here are the final exam scores of 100 students who took two semesters of a statistics sequence. In each column, there is a pair of grades. The top grade gives the final exam score of a student in the first semester course, STATS 1. The bottom grade is the exam score in STATS 2 *for the same student.* We would like to know if students who continued to the second semester improved their grades.

| STATS 1 | 53 | 58 | 67 | 46 | 63 | 42 | 35 | 35 | 35 | 48 | 53 | 51 | 34 | 52 | 31 |
|---------|----|----|----|----|----|----|----|----|----|----|----|----|----|----|----|
| STATS 2 | 63 | 46 | 79 | 58 | 60 | 77 | 84 | 71 | 65 | 80 | 61 | 65 | 80 | 58 | 68 |

| STATS 1 | 55 | 41 | 54 | 41 | 52 | 31 | 50 | 50 | 57 | 67 | 78 | 40 | 62 | 51 | 51 |
|---------|----|----|----|----|----|----|----|----|----|----|----|----|----|----|----|
| STATS 2 | 61 | 78 | 84 | 63 | 71 | 57 | 82 | 67 | 59 | 83 | 74 | 67 | 34 | 68 | 66 |

| STATS 1 | 25 | 84 | 17 | 12 | 56 | 64 | 50 | 40 | 56 | 53 | 25 | 45 | 60 | 29 | 38 |
|---------|----|----|----|----|----|----|----|----|----|----|----|----|----|----|----|
| STATS 2 | 87 | 72 | 56 | 63 | 51 | 70 | 77 | 72 | 64 | 68 | 48 | 80 | 87 | 79 | 75 |

| STATS 1 | 77 | 53 | 31 | 53 | 40 | 56 | 37 | 56 | 34 | 33 | 33 | 53 | 45 | 72 | 57 |
|---------|----|----|----|----|----|----|----|----|----|----|----|----|----|----|----|
| STATS 2 | 66 | 80 | 55 | 87 | 66 | 79 | 62 | 55 | 61 | 77 | 71 | 64 | 70 | 69 | 77 |

| STATS 1 | 74 | 67 | 29 | 65 | 73 | 34 | 47 | 48 | 63 | 57 | 40 | 28 | 72 | 57 | 15 |
|---------|----|----|----|----|----|----|----|----|----|----|----|----|----|----|----|
| STATS 2 | 65 | 77 | 72 | 78 | 73 | 82 | 76 | 71 | 53 | 71 | 70 | 55 | 68 | 46 | 54 |

| STATS 1 | 41 | 73 | 64 | 41 | 56 | 34 | 46 | 51 | 34 | 46 | 34 | 66 | 30 | 38 | 42 |
|---------|----|----|----|----|----|----|----|----|----|----|----|----|----|----|----|
| STATS 2 | 67 | 71 | 55 | 55 | 86 | 74 | 67 | 72 | 47 | 73 | 54 | 80 | 65 | 64 | 73 |

| STATS 1 | 62 | 30 | 51 | 51 | 33 | 50 | 38 | 40 | 53 | 58 |
|---------|----|----|----|----|----|----|----|----|----|----|
| STATS 2 | 82 | 72 | 66 | 76 | 80 | 77 | 72 | 48 | 75 | 69 |

We will let $u$ represent the score for STATS 1, and $v$ represent the test scores for STATS 2.

Figure 45.2.: Illustration of $p$-value for different alternate hypotheses in matched pairs testing. In each figure $t^* = \bar{x}/(s/\sqrt{n})$, where $\bar{x}$ is the mean of the sample differences; $s$ is the standard deviation of the sample differences; and $n$ is the sample size. The acutal p-value can be calculated from a t-table (e.g., table A.2 as described in chapter 40).

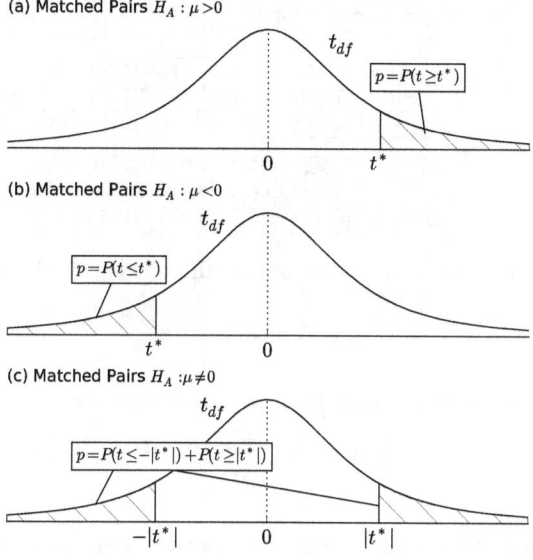

(a) Matched Pairs $H_A : \mu > 0$

$t_{df}$

$p = P(t \geq t^*)$

(b) Matched Pairs $H_A : \mu < 0$

$t_{df}$

$p = P(t \leq t^*)$

(c) Matched Pairs $H_A : \mu \neq 0$

$t_{df}$

$p = P(t \leq -|t^*|) + P(t \geq |t^*|)$

2. Make a three-column (or three-row[1]) list of your data. Label the tops of your columns (or rows) $u$ and $v$., and the third column (or row) $x = u - v$. Fill in the values (making sure they line up correctly) and calculate the differences.

(**Continuation of example 45.1.**) To save space, we write the data in rows.

| $u$ | 53 | 58 | 67 | 46 | 63 | 42 | 35 | 35 | 35 | 48 | 53 | 51 | 34 | 52 | 31 |
|-----|----|----|----|----|----|----|----|----|----|----|----|----|----|----|----|
| $v$ | 63 | 46 | 79 | 58 | 60 | 77 | 84 | 71 | 65 | 80 | 61 | 65 | 80 | 58 | 68 |
| $x = v - u$ | 10 | -12 | 12 | 12 | -3 | 35 | 49 | 36 | 30 | 32 | 8 | 14 | 46 | 6 | 37 |

---

[1]Rows may be preferred to save space, e.g., in a notebook, where the rows may need to be broken over many lines. In a spreadsheet, columns are generally easier to use as you can scroll down indefinitely as necessary.

| $u$ | 55 | 41 | 54 | 41 | 52 | 31 | 50 | 50 | 57 | 67 | 78 | 40 | 62 | 51 | 51 |
|---|---|---|---|---|---|---|---|---|---|---|---|---|---|---|---|
| $v$ | 61 | 78 | 84 | 63 | 71 | 57 | 82 | 67 | 59 | 83 | 74 | 67 | 34 | 68 | 66 |
| $x = v - u$ | 6 | 37 | 30 | 22 | 19 | 26 | 32 | 17 | 2 | 16 | -4 | 27 | -28 | 17 | 15 |

| $u$ | 25 | 84 | 17 | 12 | 56 | 64 | 50 | 40 | 56 | 53 | 25 | 45 | 60 | 29 | 38 |
|---|---|---|---|---|---|---|---|---|---|---|---|---|---|---|---|
| $v$ | 87 | 72 | 56 | 63 | 51 | 70 | 77 | 72 | 64 | 68 | 48 | 80 | 87 | 79 | 75 |
| $x = v - u$ | 62 | -12 | 39 | 51 | -5 | 6 | 27 | 32 | 8 | 15 | 23 | 35 | 27 | 50 | 37 |

| $u$ | 77 | 53 | 31 | 53 | 40 | 56 | 37 | 56 | 34 | 33 | 33 | 53 | 45 | 72 | 57 |
|---|---|---|---|---|---|---|---|---|---|---|---|---|---|---|---|
| $v$ | 66 | 80 | 55 | 87 | 66 | 79 | 62 | 55 | 61 | 77 | 71 | 64 | 70 | 69 | 77 |
| $x = v - u$ | -11 | 27 | 24 | 34 | 26 | 23 | 25 | -1 | 27 | 44 | 38 | 11 | 25 | -3 | 20 |

| $u$ | 74 | 67 | 29 | 65 | 73 | 34 | 47 | 48 | 63 | 57 | 40 | 28 | 72 | 57 | 15 |
|---|---|---|---|---|---|---|---|---|---|---|---|---|---|---|---|
| $v$ | 65 | 77 | 72 | 78 | 73 | 82 | 76 | 71 | 53 | 71 | 70 | 55 | 68 | 46 | 54 |
| $x = v - u$ | -9 | 10 | 43 | 13 | 0 | 48 | 29 | 23 | -10 | 14 | 30 | 27 | -4 | -11 | 39 |

| $u$ | 41 | 73 | 64 | 41 | 56 | 34 | 46 | 51 | 34 | 46 | 34 | 66 | 30 | 38 | 42 |
|---|---|---|---|---|---|---|---|---|---|---|---|---|---|---|---|
| $v$ | 67 | 71 | 55 | 55 | 86 | 74 | 67 | 72 | 47 | 73 | 54 | 80 | 65 | 64 | 73 |
| $x = v - u$ | 26 | -2 | -9 | 14 | 30 | 40 | 21 | 21 | 13 | 27 | 20 | 14 | 35 | 26 | 31 |

| $u$ | 62 | 30 | 51 | 51 | 33 | 50 | 38 | 40 | 53 | 58 |
|---|---|---|---|---|---|---|---|---|---|---|
| $v$ | 82 | 72 | 66 | 76 | 80 | 77 | 72 | 48 | 75 | 69 |
| $x = v - u$ | 20 | 42 | 15 | 25 | 47 | 27 | 34 | 8 | 22 | 11 |

3. Calculate the sample mean of the differences, $\bar{x}$ (chapter 13),

$$\bar{x} = \frac{x_1 + x_2 + \cdots + x_n}{n}$$

(**Continuation of example 45.1.**) Rounding to two digits,

$$\bar{x} = \frac{10 - 12 + 12 + 12 + \cdots + 22 + 11}{100} = 20.50$$

4. Calculate the sample standard deviation, $s$ (chapter 15)

$$s = \sqrt{\frac{1}{n-1} \sum (x_i - \bar{x})}$$

(**Continuation of example 45.1.**)

$$s = \sqrt{\frac{1}{99} \left( (10 - 20.5)^2 + \cdots + (11 - 20.5)^2 \right)} = 17.04$$

5. Calculate the standard error of the sample mean, (sem) $= \dfrac{s}{\sqrt{n}}$.

(**Continuation of example 45.1.**)

$$(\text{sem}) = \frac{s}{\sqrt{n}} = \frac{17.04}{\sqrt{100}} = 1.704$$

6. Calculate the **one-sample t-statistic** for $H_0: \mu = 0$,

$$t = \frac{\bar{x} - 0}{(\text{sem})} = \frac{\bar{x}}{s/\sqrt{n}}$$

(**Continuation of example 45.1.**) The t-statistic is

$$t = \frac{\bar{x} - 0}{(\text{sem})} = \frac{20.50}{1.704} = 12.0305$$

7. Determine the **degrees of freedom** $df = n - 1$, where $n$ is the sample size.

(**Continuation of example 45.1.**) Since $n = 100$, we have

$$df = n - 1 = 100 - 1 = 99.$$

8. Find the p-value (ch. 40).

(**Continuation of example 45.1.**)
To find the p-value using a table-lookup:

   a) We observe that $t = 12.0305$ and $df = 99$.
   b) We observe that there is no row 99 in table A.2, so we use the next smaller value $df$ in the table. In this case, the next smaller $df$ is 80.
   c) In row 80 in table A.2, we proceed all the way to the right, looking for t-values that bracket 12.0305. As we move to the right, the numbers increase, but the largest t-value in the row is $t = 3.195$.
   d) Following down the column from 3.195 (the largest value less than 12.0305) to the row of one-sided p-values gives $p = 0.001$.
   e) Since the p-values decrease as we move to the right, this tells us that the p-value $< 0.001$.

**Technology Example 45.2 (TI-84).** Repeat example 45.1 using the TI-84.

1. Load the data (STATS 1 and STATS 2) into two lists, e.g., **L1** and **L2**, e.g., by typing or using the data cable to load the data from a PC.
2. Use list **L3** to calculate the difference.

   a) Open the list menu by hitting [stat] [Edit] [enter].
   b) Navigate the cursor to the **L3** header. To clear out any data that is currently in **L3**, press [clear] [enter].
   c) Type [L2] [-] [L1] [enter]. (To get **L2**, use [2nd]+[2], etc.) This will calculate the difference in the entire list.

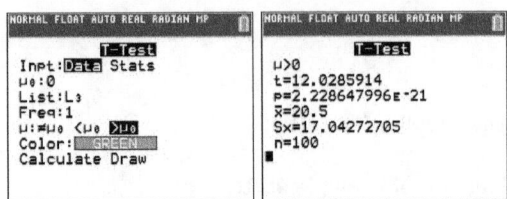

3. Select stat ≫ TESTS ≫ T-Test enter to bring up the t-test menu.

   a) For **Inpt**: select **Data**.
   b) Set $\mu_0 = 0$ (The null hypothesis is $H_0 : \mu = 0$).
   c) Set **List**: to **L3** (for the differences)
   d) Leave **Freq**: set to **1**
   e) Select $> \mu_0$ because the alternate hypothesis is $\mu > \mu_0$
   f) Scroll to **Calculate** and click **enter**

| NORMAL FLOAT AUTO REAL RADIAN MP | NORMAL FLOAT AUTO REAL RADIAN MP |
|---|---|
| **T-Test** | **T-Test** |
| Inpt:**Data** Stats | μ>0 |
| μ₀:0 | t=12.0285914 |
| List:L₃ | p=2.228647996ε-21 |
| Freq:1 | x̄=20.5 |
| μ:≠μ₀ <μ₀ **>μ₀** | Sx=17.04272705 |
| Color: **GREEN** | n=100 |
| Calculate Draw | ■ |

4. The results are displayed on the screen: $t = 12.0285914$, and the p-value for the test is $2 \times 10^{-21}$ which is vanishingly small. Therefore the test is significant.

**Technology Example 45.3 (R).** Repeat example 45.1 using R.

First, type the data into two arrays, and calculate the differences.

```
>u=c(53,58,67,46,63,42,35,35,35,48,53,51,34,52,31,
   55,41,54,41,52,31,50,50,57,67,78,40,62,51,51,25,84,
   17,12,56,64,50,40,56,53,25,45,60,29,38,77,53,31,53,
   40,56,37,56,34,33,33,53,45,72,57,74,67,29,65,73,34,
   47,48,63,57,40,28,72,57,15,41,73,64,41,56,34,46,51,
   34,46,34,66,30,38,42,62,30,51,51,33,50,38,40,53,58)
>v=c(63,46,79,58,60,77,84,71,65,80,61,65,80,58,68,
   61,78,84,63,71,57,82,67,59,83,74,67,34,68,66,87,72,
   56,63,51,70,77,72,64,68,48,80,87,79,75,66,80,55,87,
   66,79,62,55,61,77,71,64,70,69,77,65,77,72,78,73,82,
   76,71,53,71,70,55,68,46,54,67,71,55,55,86,74,67,72,
   47,73,54,80,65,64,73,82,72,66,76,80,77,72,48,75,69)
> w=v-u
```

FAQ 45. Matched Pairs Hypothesis Testing

The hypothesis test is then performed on the differences using `t.test(w)`. The default hypothesis test is for a two sided $H_A : \mu \neq 0$. To change to a one-sided test use the option `alternative="less"` or `alternative="greater"`.

```
> t.test(w)

        One Sample t-test

data:  w
t = 12.029, df = 99, p-value < 2.2e-16
alternative hypothesis: true mean is not equal to 0
95 percent confidence interval:
 17.11835 23.88165
sample estimates:
mean of x
     20.5
```

## Exercises

Perform hypothesis tests on each of the following sampling results. State $H_0$ and $H_A$.

1. The following pairs give results of exams in a freshman psychology course as (Test1, Test2).
   $(67, 85), (71, 81), (70, 62), (78, 79),$
   $(74, 71), (53, 78), (80, 73), (77, 67),$
   $(73, 62), (71, 65), (73, 76), (70, 64)$
   Does this data provide evidence of an improvement in scores?

2. Repeat the previous problem using the following data:
   $(64, 67), (74, 81), (74, 76), (70, 74),$
   $(72, 72), (66, 81), (73, 81), (65, 70),$
   $(68, 69), (73, 72), (69, 79), (71, 80),$
   $(70, 74), (68, 68), (67, 83)$
   (ans: t=3.642; one sided p-value $< .005$)

3. Fifteen men are sampled from a large group who are testing a new diet. Here are their weights in pounds (before, after):
   $(204, 202), (189, 186), (181, 179),$
   $(210, 211), (195, 195), (203, 204),$
   $(220, 221), (224, 223), (203, 197),$
   $(211, 208), (207, 204), (215, 215),$
   $(189, 187), (199, 201), (219, 216)$
   Is there evidence that the diet is effective? (ans: t-statistic = -2.39; p-value $< .025$)

# FAQ 46. How Do I Find the Degrees of Freedom ($df$) for a Difference of Means Test?

Inference for the difference of two population means $\mu_1 - \mu_2$ can be done using either confidence intervals (ch. 47) or hypothesis testing (ch. 48). Both of these procedures involve the use of the $t$ distribution and a determination of the number of **degrees of freedom**. A $t$ **distribution** with $df$ **degrees of freedom** is denoted by $t_{df}$. The process to determine the number of degrees of freedom is the same, whether you are computing a confidence interval, or doing a hypothesis test. Our notation is summarized in table 46.1.

Table 46.1.: Notation used to compare two sample means.

| | Population Mean | Sample Mean | Sample Standard Deviation | Sample Standard Error |
|---|---|---|---|---|
| Group 1 | $\mu_1$ | $\bar{x}_1$ | $s_1$ | $(\text{s.e.})_1 = s_1/\sqrt{n_1}$ |
| Group 2 | $\mu_2$ | $\bar{x}_2$ | $s_2$ | $(\text{s.e.})_2 = s_2/\sqrt{n_2}$ |

The degrees of freedom $df$ for a comparison of means $\mu_1 - \mu_2$, can be found in either of two ways, which we call the **easy method** and the **long formula method**. The long formula method is more accurate but should only be used in a computer program or calculator. The easy method should always be used for manual calculations because it is less error prone. Calculators and software typically implement the long formula method.

**Example 46.1.** Find the degrees of freedom for comparison of sample means based on the following sample statistics:

| | Sample Size | Sample Mean | Sample Standard Deviation |
|---|---|---|---|
| Group 1. | $n_1 = 17$ | $\bar{x}_1 = 76.29$ | $s_1 = 6.92$ |
| Group 2. | $n_2 = 23$ | $\bar{x}_2 = 75.87$ | $s_2 = 8.29$ |

(To be continued, below.)

**Method 1: Easy Method.** Suppose that sample 1 has size $n_1$ and sample 2 has size $n_2$. Then the number of degrees of freedom is

$$df = \text{the smaller of } (n_1 - 1) \text{ and } (n_2 - 1)$$

This method is easier to use and less likely to produce errors due to typing the wrong numbers into the calculator. However, it is slightly more conservative and will result in larger margins of error and higher $p$-values. Here is the procedure:

1. Determine $n_1$ and $n_2$ from the data. These are the sample sizes.

   (**Continuation of example 46.1.**)
   For the given example, $n_1 = 17$ and $n_2 = 23$.

2. Determine $n_1 - 1$ and $n_2 - 1$.

   (**Continuation of example 46.1.**)
   Here $n_1 - 1 = 17 - 1 = 16$, and $n_2 - 1 = 23 - 1 = 22$.

3. Then the $df$ value is the smaller of the two numbers calculated in step 2.

   (**Continuation of example 46.1.**)
   The smaller of the two numbers calculated in step 2. Hence $df = 16$.

**Method 2: Long Formula Method.** Using $s_1$, $s_2$, $n_1$ and $n_2$ we plug and chug:

$$df = \frac{\left[\dfrac{s_1^2}{n_1} + \dfrac{s_2^2}{n_2}\right]^2}{\dfrac{(s_1^2/n_1)^2}{n_1 - 1} + \dfrac{(s_2^2/n_2)^2}{n_2 - 1}}$$

If we use the shorthands $se_1 = s_1/\sqrt{n_1}$ and $se_2/\sqrt{n_2}$, and defined the **Standard Error of the Difference of the Means** (SEDM) as

$$(\text{SEDM}) = \sqrt{(\text{se})_1^2 + (\text{se})_2^2}$$

then a simpler version of the very long formula is

$$df = \frac{(\text{SEDM})^4}{\dfrac{(\text{se})_1^4}{n_1 - 1} + \dfrac{(\text{se})_2^4}{n_2 - 1}}$$

This formula is more accurate and will give tighter margins of error at the cost of more difficulty calculating. It is used by most software implementations such as calculators. It is only valid when both sample sizes are larger than 5.

1. Determine $n_1$ and $n_2$ from the data. These are the sample sizes.

**Example 46.2.** Solve example 46.1 using the long formula method. As before, we have $n_1 = 17$ and $n_2 = 23$.

2. If $(\text{se})_1$ and $(\text{se})_2$ are given (or already known), then skip to step 3. Otherwise,

   a) Find $s_1$ and $s_2$ from the sample data (see chapter 15 for a description of the standard deviation). If they are already known, continue with step 2b

   **(Continuation of example 46.2.)**
   We are given $s_1 = 6.92$ and $s_2 = 8.29$.

   b) Find the standard errors of each sample. These are given by the formulas

   $$(\text{se})_1 = \frac{s_1}{\sqrt{n_1}} \text{ and } (\text{se})_2 = \frac{s_2}{\sqrt{n_2}}$$

   **(Continuation of example 46.2.)**

   $$(\text{se})_1 = \frac{s_1}{\sqrt{n_1}} = \frac{6.92}{\sqrt{17}} = 1.68 \text{ and } (\text{se})_2 = \frac{s_2}{\sqrt{n_2}} = \frac{8.92}{\sqrt{23}} = 1.73$$

3. Find the standard error of the difference of the means. This is given by the formula

$$\text{SEDM} = \sqrt{\frac{s_1^2}{n_1} + \frac{s_2^2}{n_2}} = \sqrt{(\text{se})_1^2 + (\text{se})_2^2}$$

**(Continuation of example 46.2.)**

$$\text{SEDM} = \sqrt{(1.68)^2 + (1.73)^2} = \sqrt{5.82} = 2.41$$

4. Substitute the results of steps 1, 2b and 3 into

$$df = \frac{(\text{SEDM})^4}{\dfrac{(\text{se})_1^4}{n_1 - 1} + \dfrac{(\text{se})_2^4}{n_2 - 1}}$$

to find the degrees of freedom.

FAQ 46.  Difference of Means Degrees of Freedom

**(Continuation of example 46.2.)**

$$df = \frac{2.41^4}{\dfrac{1.68^4}{17-1} + \dfrac{1.73^4}{23-1}} = \frac{33.73}{\dfrac{7.96}{16} + \dfrac{8.96}{22}} = \frac{33.73}{.90} = 37.5$$

# Exercises

Find the two sample degrees of freedom by both methods for each of the sets of sample statistics given.

1. $n_1 = 25$, $n_2 = 17$, $s_1 = 3.7$, $s_2 = 7.3$ (ans: 16; 21.439 )

2. $n_1 = 14$, $n_2 = 16$, $s_1 = 5.6$, $s_2 = 3.9$ (ans: 13; 22.814 )
3. $n_1 = 29$, $n_2 = 72$, $s_1 = 82$, $s_2 = 85$ (ans: 28; 55.527)
4. $n_1 = 34$, $n_2 = 28$, $se_1 = 1.05$, $se_2 = 1.2$ (ans: 27; 56.887 )
5. $n_1 = 47$, $n_2 = 39$, $se_1 = .72$, $se_2 = 0.5$ (ans: 38; 83.481 )

## FAQ 47. How Do I Find a Confidence Interval for the Difference of Two Means?

Suppose the variable $x$ is normally distributed in each of two different populations. The first population has mean $\mu_1$ and standard deviation $\sigma_1$ and the second population has mean $\mu_2$ and and standard deviation $\sigma_2$. You do not know either of the the population means nor standard deviations.

Suppose you take random samples of size $n_1$ and $n_2$ from each population, respectively, and calculate the corresponding sample means $\overline{x}_1$ and $\overline{x}_2$ and standard deviations $s_1$ and $s_2$. Here we discuss how to calculate a confidence interval on the difference of the means $\mu_1 - \mu_2$ from this information (fig. 47.1).

---

### Confidence Interval on Difference of Means

If we draw independent random samples of sizes $n_1$ and $n_2$ respectively from each of two normally distributed populations, then a **level $C$ confidence interval on the difference of the means** is

$$\mu_1 - \mu_2 = (\overline{x}_1 - \overline{x}_2) \pm t^* \sqrt{\frac{s_1^2}{n_1} + \frac{s_2^2}{n_2}}$$

where $t^*$ is the critical value in a $t_{df}$-distribution such that the area under $t_{df}$ between $-t^*$ and $t^*$ is $C$ (fig. 47.2).

---

Figure 47.1.: Comparison of two means.

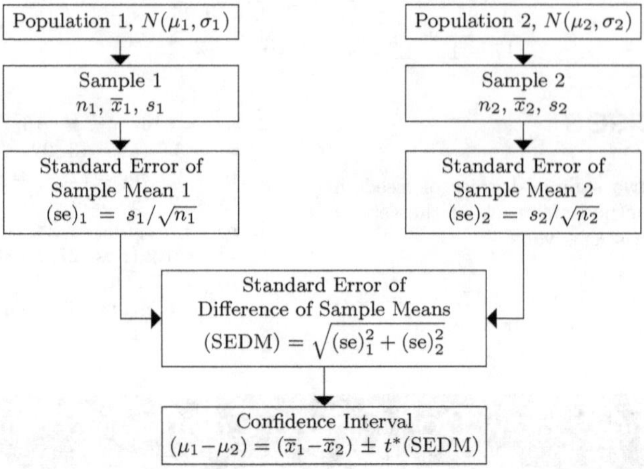

Here is the procedure to find a level $C$ confidence interval on the difference of the means $\mu_1 - \mu_2$.

1. Find the **sample means** $(\overline{x}_1, \overline{x}_2)$ (ch. 13) and **standard deviations** $(s_1, s_2)$ (ch. 15).

**Example 47.1.** Suppose we want to compare the performance of two different groups of students in calculus to see whether a refresher course in college trigonometry helped them. We will do this by finding a 95% confidence interval on the difference in grades on the final exams in samples of students taken from each population. Grades after the final exam were collected. Here is a random sample of seventeen students who took the refresher course first:

$$75, 84, 85, 85, 84, 71, 65, 62, 80, 69, 72, 79, 73, 76, 77, 81, 79$$

Here is a random sample of 23 students who did not take the refresher course:

$$66, 65, 70, 81, 81, 75, 80, 56, 67, 95, 84, 81,$$
$$78, 77, 75, 67, 75, 83, 83, 70, 79, 76, 81$$

Summary statistics are shown here:

|  | Sample Size | Sample Mean | Sample Standard Deviation |
|---|---|---|---|
| Group 1: Refresher | $n_1 = 17$ | $\overline{x}_1 = 76.29$ | $s_1 = 6.92$ |
| Group 2: No refresher | $n_2 = 23$ | $\overline{x}_2 = 75.87$ | $s_2 = 8.29$ |

2. Find the **Difference of the Sample Means** $= \overline{x}_1 - \overline{x}_2$.

FAQ 47. Difference of Means Confidence

Figure 47.2.: Definition of critical value $t^*$ for confidence level $C$.

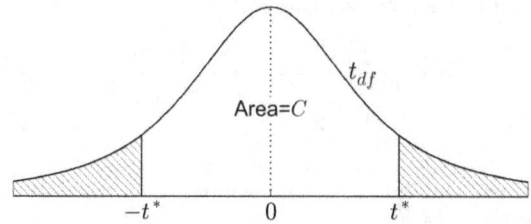

(**Continuation of example 47.1.**)

$$\text{Difference of Sample Means} = 76.29 - 75.87 = 0.42$$

3. Calculate the **Standard Error** of each sample:

$$(\text{se})_1 = \frac{s_1}{\sqrt{n_1}} \text{ and } (\text{se})_2 = \frac{s_2}{\sqrt{n_2}}$$

(**Continuation of example 47.1.**)

$$(\text{se})_1 = \frac{s_1}{\sqrt{n_1}} = \frac{6.92}{\sqrt{17}} = 1.68 \text{ and } (\text{se})_2 = \frac{s_2}{\sqrt{n_2}} = \frac{8.29}{\sqrt{23}} = 1.73$$

4. Calculate the **Standard Error of the Difference of Means**:

$$(\text{SEDM}) = \sqrt{\frac{s_1^2}{n_1} + \frac{s_2^2}{n_2}} = \sqrt{(\text{se})_1^2 + (\text{se})_2^2}$$

(**Continuation of example 47.1.**)

$$(\text{SEDM}) = \sqrt{(1.68)^2 + (1.73)^2} = \sqrt{5.82} = 2.41$$

5. Determine the **degrees of freedom** (ch. 46). Use either:

   a) **Smaller Sample Size Method:**

$$df = \text{minimum}(n_1 - 1, n_2 - 1)$$

(**Continuation of example 47.1.**)

$$df = \text{minimum}(17 - 1, 23 - 1) = \text{minimum}(16, 22) = 16$$

b) **Long Formula Method:** $df = \dfrac{(\text{SEDM})^4}{\dfrac{(\text{se})_1^4}{n_1 - 1} + \dfrac{(\text{se})_2^4}{n_2 - 1}}$

(**Continuation of example 47.1.**)

$$df = \frac{2.41^4}{\dfrac{1.68^4}{17 - 1} + \dfrac{1.73^4}{23 - 1}} = \frac{33.73}{\dfrac{7.96}{16} + \dfrac{8.96}{22}} = \frac{33.73}{.90} = 37.5$$

6. Find the critical value $t_{df}^*$ corresponding to the desired Confidence Level $C$ (ch. 39).

   (**Continuation of example 47.1.**) If you are doing the calculation by hand, use the smaller sample size $df = 16$, and table lookup. Using table A, the intersection of column for $C = 95\%$ and row 16 gives $t_{16}^* = 2.120$.

   (**Continuation of example 47.1.**) If you are using software, such as a graphing calculator, you would find $df = 37.5$ since most calculators use the long formula method. Software methods then would give $t_{37.5}^* = 2.025$.

7. Find the **Margin of Error**:

$$ME = t^* \times SEDM$$

   where (SEDM) is the standard error of the difference of means (from step 4), and $t^*$ is the value of $t_{df}^*$ (from step 6).

   (**Continuation of example 47.1.**) We had (SEDM) $= 2.41$ and either $t_{16}^* = 2.12$ (more conservative) or $t_{37.5}^* = 2.04$ (less conservative). Therefore:

   More Conservative (me) $= (2.12) \times (2.41) = 5.11$

   Less Conservative (me) $= (2.04) \times (2.41) = 4.92$

8. Find the **confidence interval**:

$$(\mu_1 - \mu_2) = (\bar{x}_1 - \bar{x}_2) \pm (\text{ME})$$

   (**Continuation of example 47.1.**) Using the results from steps 2 and 7,

   More Conservative C.I. $= 0.42 \pm 5.11 = -4.69$ to $5.53$

   Less Conservative C.I. $= 0.42 \pm 4.92 = -4.50$ to $5.34$

**Technology Example 47.2 (TI-84).** Repeat example 47.1 using a TI-84.

1. Enter each data set into a different list, such as **L1** and **L2**

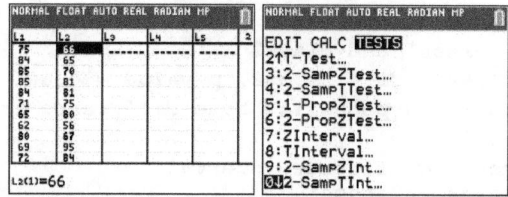

2. Select the two sample t-interval menu, $\boxed{\text{stat}} \gg \text{TESTS} \gg \boxed{\text{2-SampTInt}} \boxed{\text{enter}}$.

   a) For **Inpt:** select **Data** (because the data is in lists **L1** and **L2**
   b) For **List1:** and **List2:** select **L1** and **L2** (or whichever two lists you typed the data into).
   c) For **Freq1:** and **Freq2:** enter **1**.
   d) For **C-Level:** enter **0.95** (for 95%).
   e) For **Pooled:** select **No**.
   f) Select **Calculate** then $\boxed{\text{enter}}$.

3. The **2-SampTInt** summary window displays the results. The confidence interval is $-4.453$ to $5.3026$ and the degrees of freedom is $df = 37.36797787$. The difference with the earlier long-formula calculation is due to round-off.

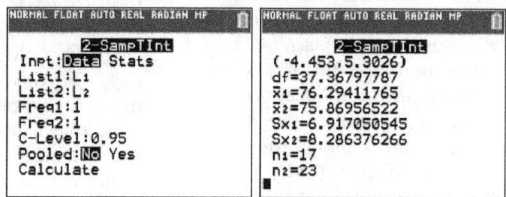

**Technology Example 47.3 (R).** Repeat example 47.1 using R.

The difference of means or two-sample t-test is performed using the **t.test** function. The syntax is:

$$\texttt{t.test(array1, array2, conf.level=value)}$$

First, enter the data into two vectors.

```
> class1=c(75,84,85,85,84,71,65,62,80,69,72,79,73,
  76,77,81,79)
> class2=c(66,65,70,81,81,75,80,56,67,95,84,81,78,
  77,75,67,75,83,83,70,79,76,81)
```

Next, perform the hypothesis using **t.test()**.

```
> t.test(class1,class2,conf.level=0.95)

        Welch Two Sample t-test

data:  class1 and class2
t = 0.17629, df = 37.368, p-value = 0.861
alternative hypothesis: true difference in means is
not equal to 0
95 percent confidence interval:
 -4.453476  5.302581
sample estimates:
mean of x mean of y
 76.29412  75.86957
```

The 95% confidence interval is $-4.453476$ to $5.402581$. The number of degrees of freedom calculated was $37.368$.

## Exercises

1. Suppose that a professor teaches two sections of Calculus, one in the early morning and one in the afternoon. She suspects that students in the afternoon are better prepared. Quiz grades for the morning class of 57 students have an average of 73 and standard deviation of 14, while quiz grades for the afternoon class of 42 students have an average of 78 with a standard deviation of 12 points. Find a 98% confidence interval on the difference $\mu_2 - \mu_1$, where $\mu_1$ is the average grade of students who take Calculus in the morning and $\mu_2$ is the average grade of students who take it in the afternoon. (Ans: -0.352 to 10.35 points)

2. A commuter measured the time it took to get to work every day for two weeks. During the first week (five days, Monday through Friday), he drove, and the average drive time was 17.4 minutes with a standard deviation of 10.3 minutes. During the second week he took the bus, and the average commute time was 18.1 minutes with a standard deviation of 4.2 minutes. Find a 95% confidence interval on the difference between the time it took him to get to work riding the bus and the time it took driving.

## FAQ 48. How Do I Do a Hypothesis Test for the Difference of Two Means?

Suppose the variable $x$ is normally distributed in each of two different populations. The first population has mean $\mu_1$ and standard deviation $\sigma_1$ and the second population has mean $\mu_2$ and and standard deviation

$\sigma_2$. You do not know either of the the population means nor standard deviations.

Suppose you take random samples of size $n_1$ and $n_2$ from each population, respectively, and calculate the corresponding sample means $\overline{x}_1$ and $\overline{x}_2$ and standard deviations $s_1$ and $s_2$. Here we discuss how to perform a hypothesis test on the difference of the means $\mu_1 - \mu_2$ from this information (fig. 48.1). In general we can test for the difference taking on any value $\mu_{1,0} - \mu_{2,0}$. The corresponding **two-sample t-statistic** follows a t-distribution with degrees freedom as described in chapter 46.

Figure 48.1.: Comparison of two means.

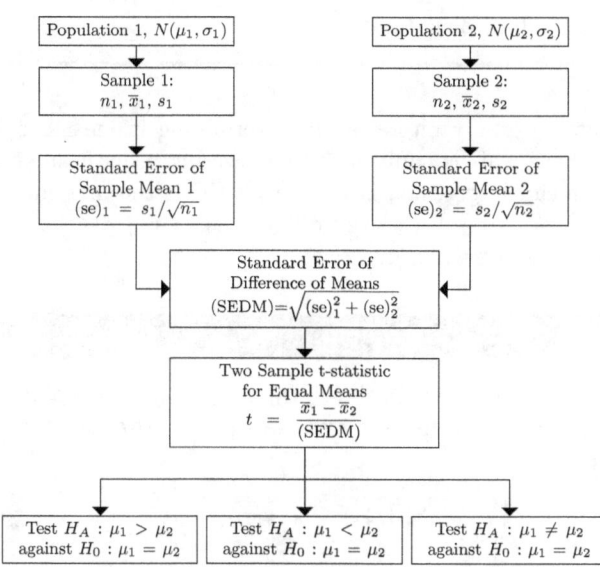

---

**Two Sample t Statistic**

**The two sample t statistic** is given by

$$t = \frac{(\overline{x}_1 - \overline{x}_2) - (\mu_1 - \mu_2)}{\text{(SEDM)}} = \frac{(\overline{x}_1 - \overline{x}_2) - (\mu_{1,0} - \mu_{2,0})}{\sqrt{s_1^2/n_1 + s_2^2/n_2}}$$

---

The two-sample t-statistic depends on standard error of the difference of the means (SEDM) that was previously in chapter 47. The SEDM is the square root of the sum of the squares of the standard errors of the means of sample 1 and sample. 2.

---

### Standard Error of the Difference of Means

The **standard error of the difference in the means** (SEDM) is

$$(\text{SEDM}) = \sqrt{\frac{s_1^2}{n_1} + \frac{s_2^2}{n_2}} = \sqrt{(\text{se})_1^2 + (\text{se})_2^2}$$

where $(\text{se})_1 = s_1/\sqrt{n_1}$ and $(\text{se})_2 = s_2/\sqrt{n_2}$ are the standard errors of the means of $x_1$ and $x_2$; and $s_1$ and $s_2$ are the corresponding sample standard deviations.

---

In general we are only interested in determining if one mean is greater than the other. This is equivalent to answering a question such as: (a) is the first mean larger, i.e., is $\mu_1 > \mu_2$?; (b) is the first mean smaller, i.e., is $\mu_1 < \mu_2$?; or (c) are they different, i.e., is $\mu_1 \neq \mu_2$? These are equivalent to testing when $\mu_{1,0} = \mu_{2,0}$.

---

### Hypothesis Test for Equal Means

To test against the hypothesis $H_0 : \mu_1 = \mu_2$ (or equivalently, $H_0 : \mu_1 - \mu_2 = 0$), use the **two sample t-statistic for equal means**:

$$t = \frac{\overline{x}_1 - \overline{x}_2}{(\text{SEDM})} = \frac{\overline{x}_1 - \overline{x}_2}{\sqrt{s_1^2/n_1 + s_2^2/n_2}}$$

---

1. Formulate the **null hypothesis**
$$H_0 : \mu_1 - \mu_2 = \mu_{1,0} - \mu_{2,0}.$$
A special case of this occurs when $\mu_{1,0} = \mu_{2,0}$ so that the null hypothesis becomes
$$H_0 : \mu_1 = \mu_2.$$
We only consider the special case here, and leave the general situation to more advanced books.

**Example 48.1.** Formulate a test to determine if twelve year old boys are

---

shorter, in general, than twelve year old girls. To do this, we let $\mu_1$ denote the average height, in centimeters of a twelve year old boy, and let $\mu_2$ denote the average height, in centimeters, of a twelve year old girl. The null hypothesis is $H_0 : \mu_1 = \mu_2$.

2. Formulate the appropriate **alternate hypothesis:**

   - $H_A : \mu_1 > \mu_2$ (one sided hypothesis)
   - $H_A : \mu_1 < \mu_2$ (one sided hypothesis)
   - $H_A : \mu_1 \neq \mu_2$ (two sided hypothesis)

   (**Continuation of example 48.1.**) If boys are shorter, then we expect $\mu_1$ to be smaller than $\mu_2$. So the alternate hypothesis is $H_A : \mu_1 < \mu_2$.

3. **Determine $\alpha$** (the significance level, ch. 34). Typical values are 5%, obtained by setting $\alpha = 0.05$, or smaller.

   (**Continuation of example 48.1.**) We will choose $\alpha = 0.05$.

4. Obtain **random samples** from each population. Denote the sample sizes by $n_1$ and $n_2$.

   (**Continuation of example 48.1.**)   Here are samples of the heights of $n_1 = 17$ twelve year old boys in centimeters:
   $$151.79,\ 150.66,\ 150.22,\ 153.59,\ 156.55,\ 153.12,$$
   $$142.39,\ 148.83,\ 141.36,\ 142.15,\ 147.76,\ 147.81,$$
   $$149.54,\ 149.39,\ 151.15,\ 145.01,\ \ 152.36$$
   Here are the heights fifteen twelve year old girls ($n_2 = 15$) in centimeters:
   $$153.66,\ 145.56,\ 149.43,\ 151.29,\ 148.70,$$
   $$149.55,\ 155.53,\ 152.94,\ 151.24,\ 152.84,$$
   $$148.20,\ 154.63,\ 150.93,\ 153.65,\ \ 152.57$$

5. Calculate the **sample means** $\overline{x}_1$ and $\overline{x}_2$ (ch. 13).

   (**Continuation of example 48.1.**) $\overline{x}_1 = 149.04$ and $\overline{x}_2 = 151.38$.

6. Calculate the **sample standard deviations** $s_1$ and $s_2$ from the sample data (ch. 15).

   (**Continuation of example 48.1.**)   $s_1 = 4.27$ and $s_2 = 2.71$.

7. Calculate the **sample standard errors**

---

$$(\text{se})_1 = s_1/\sqrt{n_1}, \qquad (\text{se})_2 = s_2/\sqrt{n_2}$$

(**Continuation of example 48.1.**)
$$(\text{se})_1 = s_1/\sqrt{n_1} = 4.27/\sqrt{17} = 1.036$$
$$(\text{se})_2 = s_2/\sqrt{n_2} = 2.71/\sqrt{15} = 0.6997$$

8. Calculate the **standard error of the difference of means**,
$$(\text{SEDM}) = \sqrt{(\text{se})_1^2 + (\text{se})_2^2}$$

(**Continuation of example 48.1.**)
$$(\text{SEDM}) = \sqrt{(\text{se})_1^2 + (\text{se})_2^2} = \sqrt{(1.036)^2 + (0.6997)^2} = 1.25$$

9. Calculate the **two-sample t-statistic** $t = \dfrac{\bar{x}_1 - \bar{x}_2}{(\text{SEDM})}$

(**Continuation of example 48.1.**)
$$t = \frac{\bar{x}_1 - \bar{x}_2}{(\text{SEDM})} = \frac{149.04 - 151.38}{1.25} = -1.872$$

10. Determine the **degrees of freedom** (ch. 46).

(**Continuation of example 48.1.**) Using the smaller sample size method, since $n_1 = 17$ and $n_2 = 15$, there are 14 degrees of freedom.

11. **Find the p-value** (ch. 40).

(**Continuation of example 48.1.**) Using table A.2, for $df = 14$, with $t = 1.872$ gives $1.761 < t < 2.145$. Following these columns to the bottom of the table, the one-sided test row gives bounds of $.025 < $ p-value $ < .05$

12. **Compare the p-value to $\alpha$.** If $p \leqslant \alpha$, the test is significant at level $\alpha$. We generally conclude, in this case, that there is sufficient evidence to reject $H_0$ and accept $H_A$. If $p$ is larger than $\alpha$ the test is inconclusive.

(**Continuation of example 48.1.**) Since $p < 0.05$ and $\alpha = 0.05$ we conclude that $p \leqslant \alpha$. Therefore there is conclusive evidence at the five percent significance level that twelve year old boys are shorter than twelve year old girls.

**Technology Example 48.2 (TI-84).** Repeat example 48.1 using a TI-84.

---

1. Put the each data sample into a separate list, such as **L1** and **L1**.
2. Select the two sample T-Test mode, stat $\gg$ TESTS $\gg$ 2-SampTTest enter.

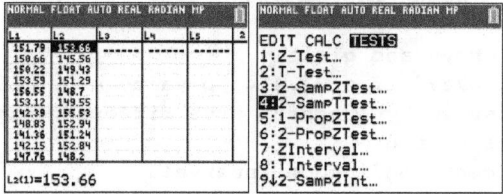

   a) For **Inpt**: select **Data** (because the data is in the two lists).
   b) For **List1**: and **List2**: enter **L1** and **L2** (or whichever lists you put the data into.)
   c) For **Freq1**: and **Freq2**: enter **1**.
   d) Select the test $< \mu_2$ to test $H_A : \mu_1 < \mu_2$.
   e) For **Pooled**: select **No**
   f) Navigate to **Calculate** and click enter.

3. The **2-SampTTest** summary window will display the results for the test $H_A : \mu_1 < \mu_2$. The $t$ value is $-1.874686218$, the $p$ value is $0.0357640916$, and the degrees of freedom are $df = 27.41363162$. This is consistent with the manual calculation performed in example 48.1.

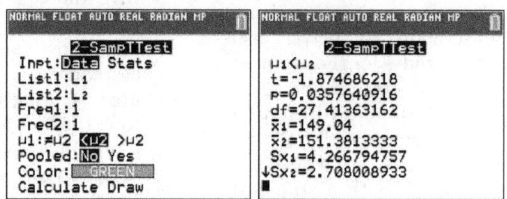

**Technology Example 48.3 (R).** Repeat example 48.1 in R.

We can perform a two-sample t-test with **t.test**.

```
> boys=c(151.79,150.66,150.22,153.59,156.55,153.12,
    142.39, 148.83, 141.36, 142.15, 147.76, 147.81,
    149.54, 149.39, 151.15, 145.01, 152.36)
> girls=c(153.66,145.56,149.43,151.29,148.70,149.55,
    155.53,152.94, 151.24, 152.84, 148.20, 154.63, 1
    50.93, 153.65, 152.57)
> t.test(boys,girls,conf.level=0.95,
    alternative="less")
```

```
            Welch Two Sample t-test

data:  boys and girls
t = -1.8747, df = 27.414, p-value = 0.03576
alternative hypothesis: true difference in means
is less than 0
95 percent confidence interval:
        -Inf -0.2152013
sample estimates:
mean of x mean of y
 149.0400  151.3813
```

The $p$-value is 0.03576, consistent with the other methods.

## Exercises

1. Suppose that a professor teaches two sections of Calculus, one in the early morning and one in the afternoon. She suspects that students in the afternoon are better prepared. Quiz grades for the morning class of 57 students have an average of 73 and standard deviation of 14, while quiz grades for the afternoon class of 42 students have an average of 78 with a standard deviation of 12 points. Does this support the conjecture that the afternoon class is better prepared? Formulate a hypothesis test, state your confidence level (*alpha*), compute the p value, and make a conclusion.

2. A commuter measured the time it took to get to work every day for two weeks. During the first week (five days, Monday through Friday), he drove, and the average drive time was 17.4 minutes with a standard deviation of 10.3 minutes. During the second week he took the bus, and the average commute time was 18.1 minutes with a standard deviation of 4.2 minutes. Formulate and evaluate a hypothesis to determine if this observation supports the conjecture that driving a car is faster than riding the bus.

# FAQ 49. How Do I Find A Confidence Interval on a Proportion?

In general we do not know the proportion of an item in a population and we will want to estimate it with sampling.

In this chapter we will discuss the procedure on estimating a confidence interval for the true population proportion (fig. 49.1). In chapter 50 we will discuss hypothesis tests on proportions, and in chapters 52 and 53 we will discuss the difference of proportions.

---

**Population Proportion**

The **population proportion of** $A$ **is**

$$p = \frac{\text{Number of occurances of } A \text{ in the Population}}{\text{Size of Population}}$$

---

Figure 49.1.: Confidence interval for proportions.

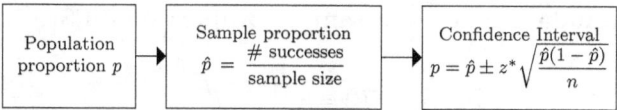

If a random sample of size $n$ of a variable $x$ is drawn from a population, then the number of times that $x$ takes on the value $A$ in the sample is called the **number of successes**. The sample proportion is $\hat{p} = x/n$.

---

**Sample Proportion**

The **sample proportion** of a variable is

$$\hat{p} = \frac{x}{n}$$

where $x$ is the number of successes, and $n$ is the sample size. Here $\hat{p}$ is read as "p hat."

---

**Example 49.1.** Suppose you are interested in determining the proportion of orange candies that are typically put into a bag of m&m's. To do this you buy several bags of candy and mix them together in hopes of getting a reasonably large and well mixed random sample. Then you count all the m&m's in the mix. Suppose that there are $n = 347$ total pieces of candy. Next, you separate out the orange candy, count them, and find a total of $x = 82$. We call these 82 orange m&m's the number of successes. The sample proportion is then

$$\hat{p} = \frac{\text{successes}}{\text{sample size}} = \frac{x}{n} = \frac{82}{347} = 0.2363$$

---

A proportion is always a number between 0 and 1 (inclusive). However, proportions, are sometimes expressed in writing as percentages. The sample proportion $\hat{p} = 0.2363$ of orange m&m's found in example 49.1 is equivalent to the following statement: "The sample contains 23.63% orange colored candy." **Note, however, that of the equations we write down for proportions are in terms of numbers between 0 and 1, and not in terms of percentages** (fig 49.2). Thus, for example, if we calculate a margin of error with to within, say, $\pm 3$ percentage points, this means that the margin of error on $p$ is really 0.03.

---

### Standard Error of Sample Proportion

The **standard error of a sample proportion** (SESP) is

$$(\text{SESP}) = \sqrt{\frac{\hat{p}(1 - \hat{p})}{n}}$$

---

Figure 49.2.: A proportion is a number between 0 and 1; a percentage is a number between 0 and 100. To convert multiply or divide by 100.

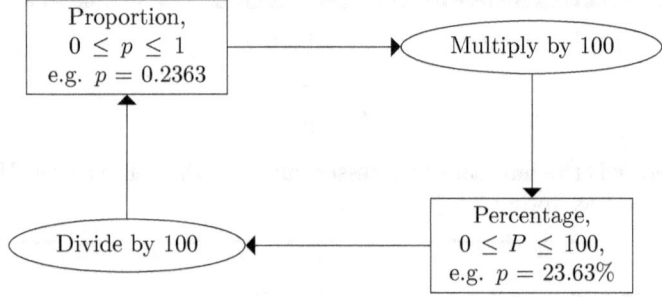

The confidence interval depends on the standard error of the sample proportion, which is given by the formula $(\text{SESP}) = \sqrt{\hat{p}(1 - \hat{p})/n}$, where $\hat{p}$ is the sample proportion. The margin of error for a level $C$ confidence interval is $z^* \times (\text{SESP})$, where $z^*$ is the critical number defined so as to give a central area equal to C, under a standard normal distribution, between $-z^*$ and $z^*$ (fig. 49.3). In terms of the standard error of the

---

sample proportion, the confidence interval on a proportion is then

$$p = \hat{p} \pm z^*(\text{SESP})$$

---

### Confidence Interval for a Proportion

If a random sample of size $n$ is drawn from a normally distributed population then a level $C$ confidence interval for the population proportion $p$ of some variable is

$$p = \hat{p} \pm z^*\sqrt{\frac{\hat{p}(1-\hat{p})}{n}}$$

where $z^*$ is the critical value (see fig. 49.3). This estimate is generally valid when both $n\hat{p} \geqslant 15$ and $n(1-\hat{p}) \geqslant 15$.

---

Figure 49.3.: The critical number $z^*$ is defined such that the central area between $-z^*$ and $z^*$, under the curve of a standard normal distribution, is equal to $C$.

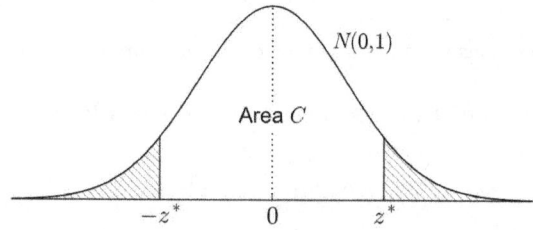

To find a confidence interval for a proportion:

1. Find the **sample proportion** $\hat{p} = \dfrac{(\text{count of successes})}{(\text{total count of sample})}$.

   Note: If $\hat{p}$ is very close to zero or very close to one, modify this estimate as follows:

   $$\hat{p} = \frac{(\text{count of successes}) + 2}{(\text{total count of sample}) + 4}$$

---

**Example 49.2.** Suppose you are having a school party and the school color is orange. You want to have a lot of orange m&m's at the party, but the local store only sells mixed colors. The m&m company will not tell you the distribution of orange m&m's in the mixed bags. You buy five bags and mix them together to estimate the proportion, so that you know how many bags you need to buy. Each bag contains 75 candies. You count the orange candies and compute a 95 confidence interval on the proportion $p$.

You find a total of 81 orange candies all together in the five bags.

The total number of success is $x = 81$. The total sample size is $n = 5 \times 75 = 375$ (each bag has 75 candies and you used five bags). Therefore

$$\hat{p} = \frac{x}{n} = \frac{81}{375} = 0.216.$$

2. **Verify that the conditions** for a large sample are met, i.e., that both $n\hat{p} \geqslant 15$ and $n(1 - \hat{p}) \geqslant 15$.

**(Continuation of example 49.2.)** Since

$$n\hat{p} = 375 \times 0.216 = 81 \geqslant 15$$
$$n(1 - \hat{p}) = 375 \times (1 - .216) = 294 \geqslant 15$$

both conditions for a large sample confidence interval are met.

3. Calculate the **standard error of the sample proportion**:

$$(\text{SESP}) = \sqrt{\frac{\hat{p}(1 - \hat{p})}{n}}$$

**(Continuation of example 49.2.)** The SESP for the orange m&m's is

$$(\text{SESP}) = \sqrt{\frac{0.216(1 - 0.216)}{375}} = 0.021.$$

4. Determine the **critical value** $z^*$ corresponding to the confidence level $C$ (e.g., following one of the methods described in chapter 29).

**(Continuation of example 49.2.)**
For a 95% confidence interval, we use $z^* = 1.960$.

5. Calculate the **margin of error**: $(\text{me}) = z^* \times (\text{SESP})$

(**Continuation of example 49.2.**)
$$(\text{me}) = z^* \times (\text{SESP}) = (1.960) \times (0.021) = 0.041.$$

6. The level $C$ **confidence interval** on $p$ is: $p = \hat{p} \pm (\text{me})$

(**Continuation of example 49.2.**)
$$p = \hat{p} \pm (\text{me}) = 0.216 \pm 0.041 \text{ or } 0.175 \text{ to } 0.257.$$
We believe that $p$ is between 17.5% and 25.7% with 95% confidence.

**Technology Example 49.3 (TI-84).** Repeat example 49.2 using a TI-84.

1. Select the **1-PropZInt** menu: [stat] ⟫ TESTS ⟫ 1-PropZint [enter].
   a) For **x:** enter **81** (the number of successes);
   b) For **n:** enter **375** (the total sample size);
   c) For **C-Level:** enter **0.95** (for a 95% confidence level);
   d) Navigate to **Calculate** and press [enter]

2. The solution appears on the **1-PropZInt** results screen. The confidence interval is shown as $(0.01743, 0.2577)$ and $\hat{p} = 0.216$. These results are consistent with example 49.2.

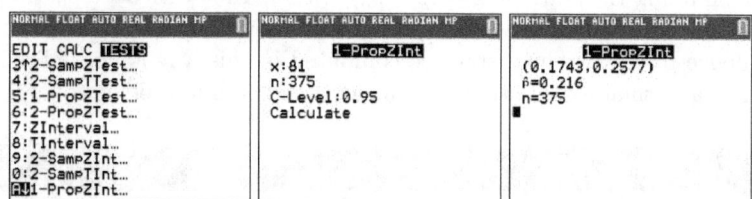

**Technology Example 49.4 (R).** Repeat example 49.2 in R.

To find $z^*$ we need to use **qnorm((1+C)/2**.

```
> successes=81
> phat=successes/samplesize
> sem=sqrt(phat*(1-phat)/samplesize)
> zstar=qnorm((1+.95)/2)
> me=zstar*sem
> phat-me
[1] 0.1743498
> phat+me
[1] 0.2576502
```

The confidence interval is 0.1743498 to 0.2576502.

# Exercises

Find confidence intervals on the following proportions.

1. $n = 25$, C=95%, number of successes = 14
2. $n = 37$, $\hat{p} = .29$, C=98%
3. $n = 12,427$, $\hat{p} = 0.52$, C=95%
4. The proportion of red milk chocolate m&m's that are produced. Using a small bag that contains 31 candies as your simple random sample you find 7 reds. Use C=98%.
5. While you were eating lunch in front of the school library the other day, you observed that 37 of the 48 students who walked by were wearing some sort of earphones or headsets. Find a 96% confidence interval on the actual proportion of students who wear headsets when walking between classes.

# FAQ 50. How Do I Do A Hypothesis Test on a Proportion?

We can perform a hypothesis test against the hypothesis $H_0 : p = p_0$ by taking a random sample of size $n$, computing the sample proportion $\hat{p}$, and examining the statistic $z = (\hat{p} - p_0)/(\text{SEP})$. Here SEP $= \sqrt{p_0(1 - p_0)/n}$ is the standard error of the proportion, assuming that $H_0$ is true (and hence $p = p_0$). The z statistic computed in this way is normalized, and has a normal distribution with mean 0 and standard deviation 1.

---

**Hypothesis Test for Proportion**

To test against the hypothesis $H_0 : p = p_0$ use the statistic

$$z = \frac{\hat{p} - p_0}{\sqrt{\dfrac{p_0(1 - p_0)}{n}}}$$

on a simple random sample of size $n$. The z statistic has a normal distribution with mean 0 and standard deviation 1 for suitably large $n$. This is generally true for large samples such that $n \geqslant 20$ and both $n\hat{p} \geqslant 10$ and $n(1 - \hat{p}) \geqslant 10$.

---

1. State the null ($H_0$) and alternative ($H_A$) hypotheses. The null

hypothesis is
$$H_0 : p = p_0$$
for some value of $p_0$. The only possible alternative hypotheses:
$$H_A : p > p_0$$
$$H_A : p < p_0$$
$$H_A : p \neq p_0$$

The first two are one-sided hypotheses and the third is a two-sided hypothesis.

**Example 50.1.** In a survey of 1250 American voting age adults by a major foundation in October 2016, 54% said that they were following the Zika epidemic fairly closely or very closely in the news, and the remaining 46% said that they were not following it very closely or not following it at all. Is this evidence that Americans of voting age were concerned about Zika in October 2016? Let $p$ be the proportion of adults who are concerned, let $p_0 = 0.5$. If $p > p_0$, then people are concerned, and if $p < p_0$, people are unconcerned. In fact, we are told that $\hat{p} = 0.54$. We use the one-sided hypothesis test $H_0: p = 0.5$, $H_A: p > 0.5$.

2. Define what it means for your results to be **statistically significant**. In other words (ch. 34), specify a number $\alpha$ such that

   ▸ If you assume $H_0$ is true; and
   ▸ You calculate that $p \leqslant \alpha$; then
   ▸ You are willing to reject $H_0$

   The second bullet point means that the observation or data is extremely unlikely to occur by random chance if $H_0$ is true, and therefore must be meaningful.

   **(Continuation of example 50.1.)** We will use a one percent confidence level, i.e, set $\alpha = 0.01$.

3. Find the **sample proportion**
   $$\hat{p} = \frac{(\text{count of successes})}{(\text{total count of sample})}$$

   Note: If $\hat{p}$ is very close to zero or very close to one, modify this estimate as follows:
   $$\hat{p} = \frac{(\text{count of successes}) + 2}{(\text{total count of sample}) + 4}$$

(**Continuation of example 50.1.**) We are given that $\hat{p} = 0.54$.

4. **Verify that the conditions for a large sample are met, i.e., that $n \geqslant 20$ with both $n\hat{p} \geqslant 10$ and $n(1 - \hat{p}) \geqslant 10$.**

(**Continuation of example 50.1.**) Since $n = 1250$ and $\hat{p} = 0.54$, we have $n\hat{p} = (1250)(0.54) = 654$ and $n(1 - \hat{p}) = (1250)(1 - .54) = 575$. Both exceed 10, so the conditions are met.

5. Calculate the **standard error of the proportion**, assuming that $H_0$ is true:

$$(\text{SEP}) = \sqrt{\frac{p_0(1 - p_0)}{n}}$$

(**Continuation of example 50.1.**) Using $n = 1250$ and $p_0 = 0.5$,

$$(\text{SEP}) = \sqrt{\frac{p_0(1 - p_0)}{n}} = \sqrt{\frac{0.5(1 - 0.5)}{1250}} = 0.014$$

6. Calculate the **z-statistic** $z = \dfrac{\hat{p} - p_0}{(\text{SEP})}$.

(**Continuation of example 50.1.**) Using $(\text{SEM}) = 0.014$, $\hat{p} = 0.54$ and $p_0 = 0.5$,

$$z = \frac{\hat{p} - p_0}{(\text{SEP})} = \frac{0.54 - 0.5}{0.014} = 2.83$$

7. Find the *p-value* (ch. 33; table A.1; table 50.1).

(**Continuation of example 50.1.**) The table value listed for $z = 2.83$ in table A.1 is 0.9976. Since we are doing a one-sided test on a right tail p-value, we calculate that:

$$\text{p-value} = 1 - \text{TableValue}(z) = 1 - 0.9976 = 0.0024$$

8. **Compare the p-value with $\alpha$. If the p-value $\leqslant \alpha$, reject $H_0$. If the p-value $> \alpha$, do not reject $H_0$.**

(**Continuation of example 50.1.**) Since $\alpha = 0.01$ we conclude that

$$\text{p-value} = 0.0024 < 0.01 = \alpha$$

Since the p-value is smaller than $\alpha$ the data provides convincing evidence that we should reject $H_0$ and accept $H_A$. In other words, it supports the conjecture (at the 2% confidence level) that Americans were concerned about Zika in October 2016.

Table 50.1.: Definition of tail probabilities that give p-values for the three possible alternate hypotheses when testing for a sample proportion against $H_0$ : $p = p_0$.

| Alternative Hypothesis | $p$ value |
|:---:|:---:|
| $H_A : p > p_0$ | $P(Z < z)$ |
| $H_A : p < p_0$ | $P(Z > z)$ |
| $H_A : p \neq p_0$ | $2P(Z > |z|)$ |

**Technology Example 50.2 (TI-84).** Repeat example 50.1 using a TI-84.

1. Open the **1-PropZTest** menu by stat ⟫ TESTS ⟫ 1-PropZTest enter .

   a) For **p0** type in **0.5** (because we are testing against $H_0 : \mu = 0.5$);

   b) For **x** type in the number of successes; since we are given $n = 1250$ and $54\%$, we should type in **0.54 * 1250** here;

   c) For **n** type in **1250** (the sample size);

   d) For **prop** select $> p_0$ (to test the alternative hypothesis $H_A : \mu > 0.5$);

   e) Navigate to **Calculate** and press enter

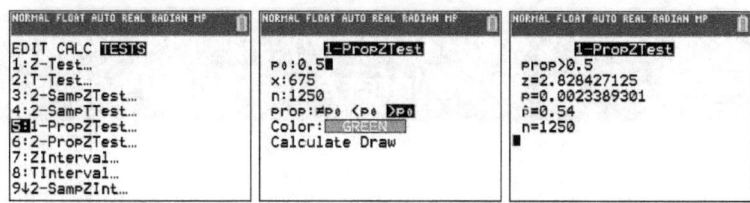

2. The results are then immediately displayed on the **1-PropZTest** output screen. The z statistic is shown to be 2.828427125, and the p value is 0.0023389301. This result is consistent with example 50.1.

**Technology Example 50.3 (R).** Repeat example 50.1 using R.

There is no specific function for this but we can calculate the z-value and then look up the p value using **pnorm**, which gives the area to the left of its argument. To the get the area to the right of **z** we use **1-pnormz**.

```
> n=1250
> phat=0.54
> p0=0.5
> sep=sqrt(p0*(1-p0)/n)
> z=(phat-p0)/sep
```

```
> pvalue = 1-pnorm(z)
> pvalue
[1] 0.002338867
```

# Exercises

1. The m&m company published in 2011 that 23% of milk chocolate m&m's were red. You purchase a bag that contains 300 candies and find that 75 are red. Thus

$$\hat{p} = \frac{\text{number of reds}}{\text{total number of candies}}$$

Is this evidence that that more than 23% are red?

2. Flip a coin 25 times and count the number of heads. Let

$$\hat{p} = \frac{\text{number of heads}}{\text{number of coin flips}}$$

Evaluate the alternate hypothesis $H_A : p \neq 0.5$ against $H_0 : p = 0.5$.

3. In a survey of 746 American parents in southern California, 368 said that all children should probably get immunized before starting pre-school or should definitely get immunized before starting pre-school, while the remainder said that children either should probably not get immunized until they were older or should definitely not get immunized until they were older. Is this evidence that southern Californians prefer later immunization over early immunization?

# FAQ 51. How Big Does My Sample Size Have to Be to Do Inference on A Proportion?

Take a random sample of size $n$ and compute a sample proportion $\hat{p}$. The margin of error (me) for a level C confidence is (ch. 49)

$$(\text{me}) = z^* \sqrt{\frac{\hat{p}(1 - \hat{p})}{n}}$$

Here $z^*$ is the critical value in a standard normal distribution such that the central area between $-z^*$ and $z^*$ is C. Solving for $n$ gives

$$n = \left[ \frac{z^*}{(\text{me})} \right]^2 \hat{p}(1 - \hat{p})$$

This equation presents something of a conundrum:

▸ To determine $n$ (the sample size) we need to know $\hat{p}$.
▸ To determine a value $\hat{p}$ (the sample proportion) we need a sample, which means we need to know $n$.

We need $\hat{p}$ to get $n$ and you need $n$ to get $\hat{p}$.

To solve this, we scratch our tails a bit and pretend we know what we are doing by guessing: we replace $\hat{p}$ with $p^*$, our **best guess** of $\hat{p}$.

The following approaches are typically used:

- Any preliminary results from previous experiments or surveys.
- Take a small survey or do a preliminary experiment to get a preliminary value for $p^*$.[1]
- As a last result, set $p^* = 0.5$. This will over-estimate the margin of error and give the largest value of $n$, so it will err on the conservative side.

---

### Sample Size to Estimate a Proportion

The minimum sample size required to estimate a proportion with a confidence level C and margin of error (me) is

$$n = \left[ \frac{z^*}{(\text{me})} \right]^2 p^*(1 - p^*)$$

where $p^*$ is the best guess at the true proportion (or $1/2$), and $z^*$ is the critical value corresponding to a confidence level C, defined such that the central area between $-z^*$ and $z^*$ under a standard normal distribution is equal to C.

---

## Exercises

Find the minimum sample size required to find confidence intervals with the specified margins of error at the given levels of confidence.

1. 95% confidence interval, $\pm 1\%$ margin of error, $p^* = .2$
2. 90% confidence interval, $\pm 3\%$ margin of error, $p^*$ unknown.
3. 98% confidence interval, $\pm 0.01$ margin of error, $p^* = 0.75$
4. 95% confidence interval, 0.01 margin of error, $p^*$ unknown.

5. At a graduation party the host has a large mixed bottle of blue and yellow candies that represent the school colors. As a party game, guests are asked to estimate the proportion of candies that are blue. You are allowed to shake up the bottle and take one look through the top to estimate the proportion. When you get your chance you see a total of 7 candies that are all blue, and no yellow candies. You know that there are yellow candies in the bottle, so an es-

---

[1] If the sample is too small you run a higher risk of getting a value that is near 1 or 0, and you should calculate $\hat{p} = (\text{number of successes} + 2)/(\text{total count} + 4)$.

timate of $\hat{p} = 1$ is incorrect. Make a correction by adding in two false success and two false failures. How large of a sample would you need to estimate the proportion of candy that are blue? Assume you want a 95% confidence interval with $\pm 0.01$ margin of error. Repeat the estimate using $p^* = 0.5$.

# FAQ 52. How Do I Find A Confidence Interval on a Difference in Proportions?

Suppose we take independent samples from two distinct populations, and calculate the sample proportion $\hat{p}$ in each sample. We now want to find a confidence interval on the difference in proportions between the two populations.

**Example 52.1.** Are there different proportions of red candies in milk chocolate m&m's and peanut m&m's? We sample a bag of 307 milk chocolate m&m's and a bag of 243 peanut m&m's. We find 71 milk chocolate reds and 48 red peanut candy. What is the difference in proportions $p_{\text{milk chocolate}} - p_{\text{peanut}}$?

Figure 52.1.: Confidence interval on a difference of proportions. The samples are taken from different populatios and must be independent.

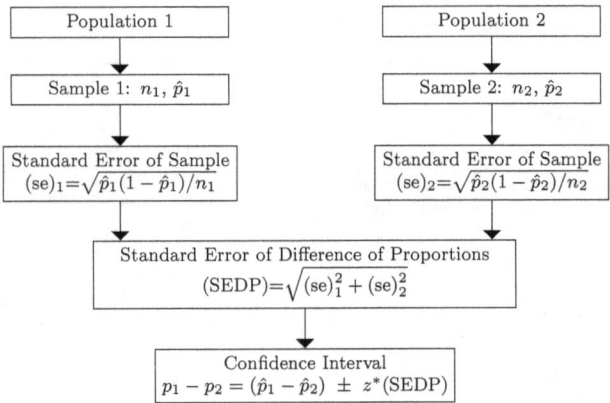

The procedure for determining the confidence interval is illustrated in fig. 52.1. The standard error of the sample proportions are $\sqrt{\hat{p}_1(1 - \hat{p}_1)/n_1}$ and $\sqrt{\hat{p}_2(1 - \hat{p}_2)/n_2}$, respectively (ch. 49). The standard error of the difference in proportions, (SEDP), is then the square root of the sum of the squares standard errors of the sample proportions: $(\text{SEDP})=\sqrt{(\text{se})_1^2 + (\text{se})_2^2}$. The difference is described by a normal distribution, for both $n$ sufficiently large. A level $C$ confidence interval has margin of error $z^* \times (\text{SEDP})$, where $z^*$ is the critical number defined such that the central area under a standard normal distribution between $-z^*$ and $z^*$ is $C$.

---

### Confidence Interval: Difference in Proportions

A level $C$ **confidence interval on the difference between two proportions** $p_1 - p_2$ is given by

$$p_1 - p_2 = (\hat{p}_1 - \hat{p}_2) \pm z^* \sqrt{\frac{\hat{p}_1(1 - \hat{p}_1)}{n_1} + \frac{\hat{p}_2(1 - \hat{p}_2)}{n_2}}$$

where $z^*$ is the critical value such that the central area in a standard normal distribution with $N(0, 1)$ gives probability $C = P(-z^* \leqslant z \leqslant z^*)$. Here $n_1$ and $n_2$ are the sizes of simple random samples from two independent, normally distributed populations, and $\hat{p}_1$ and $\hat{p}_2$ are the sample proportions.

---

1. **Obtain independent random samples** of size $n_1$ and $n_2$ from two populations and calculate the **sample proportions** $\hat{p}_1$ and $\hat{p}_2$. The sample proportions are:

$$\hat{p}_1 = \frac{(\text{number of successes in sample 1})}{n_1}$$
$$\hat{p}_2 = \frac{(\text{number of successes in sample 2})}{n_2}$$

**Example 52.2.** To determine the difference in the proportions of red candy in milk chocolate m&m's and peanut m&m's you purchase one medium size bag of each (about three-quarters of a pound) at the local market to use as your random sample. The bag of peanut candy contains $n_1 = 147$ pieces and the bag of plain milk chocolate contains $n_2 = 374$ pieces. You then count the number of red candy in each bag and find 17 red peanut candy and 52 plain red candy. Thus

$$\hat{p}_1 = \frac{17}{147} = 0.1156$$
$$\hat{p}_2 = \frac{52}{374} = 0.1390$$

Your goal is to calculate a 95% confidence interval on the difference of the population proportions $p_1 - p_2$.

2. **Find the difference in the sample proportions $\hat{p}_1 - \hat{p}_2$.**

    **(Continuation of example 52.2.)**

    $$\hat{p}_1 - \hat{p}_2 = 0.1156 - 0.1390 = -0.0234$$

3. Calculate the **standard error of the proportion** for each sample,

    $$(\text{se})_1 = \sqrt{\hat{p}_1(1 - \hat{p}_1)/n_1} \text{ and } (\text{se})_2 = \sqrt{\hat{p}_2(1 - \hat{p}_2)/n_2}.$$

    **(Continuation of example 52.2.)**

    $$(\text{se})_1 = \sqrt{\hat{p}_1(1 - \hat{p}_1)/n_1} = \sqrt{0.1156(1 - 0.1156)/147} = 0.0264$$
    $$(\text{se})_2 = \sqrt{\hat{p}_2(1 - \hat{p}_2)/n_2} = \sqrt{0.1390(1 - 0.1390)/374} = 0.0179$$

4. Find the **standard error of the difference of proportions,**

    $$(\text{SEDP}) = \sqrt{(\text{se})_1^2 + (\text{se})_2^2}$$

    **(Continuation of example 52.2.)**

    $$(\text{SEDP}) = \sqrt{(\text{se})_1^2 + (\text{se})_2^2} = \sqrt{.0264^2 + 0.0179^2} = 0.0319.$$

5. Find the **critical value** $z^*$ for a level $C$ confidence interval.

    **(Continuation of example 52.2.)** For 95%, $z^* = 1.96$.

6. Calculate the bounds for the confidence interval:

    $$(p_1 - p_2) = (\hat{p}_1 - \hat{p}_2) \pm z^*(\text{SEDP})$$

    **(Continuation of example 52.2.)**

    $$\begin{aligned}(p_1 - p_2) &= (\hat{p}_1 - \hat{p}_2) \pm z^*(\text{SEDP})\\ &= -0.0234 \pm (1.96)(0.0319)\\ &= -0.0234 \pm 0.0625\\ &= -0.0859 \text{ to } 0.0391\end{aligned}$$

The confidence interval is $-8.59\%$ to $3.91\%$, with a center at $-2.34\%$.

---

FAQ 52. Difference of Proportions, Confidence Interval

7. **Interpret** the result.

(**Continuation of example 52.2.**)  Writing the confidence interval as an inequality, in percent, gives

$$-8.59\% < p_1 - p_2 < 3.91\%$$

Adding $p_2$ to both sides of the equation we have

$$p_2 - 8.59\% < p_1 < p_2 + 3.91\%$$

Since $p_1$ is the proportion (represented here by a percent) of red plain candy, and $p_2$ is the proportion of red peanut candy:

$$(\% \text{ of red peanut}) - 8.59 < (\% \text{ of red plain})$$
$$< (\% \text{ of red peanut}) + 3.91$$

As to the prediction of "which is more?" our mean difference of proportions was

$$\hat{p}_1 - \hat{p}_2 = (\text{proportion of red plain candy})$$
$$- (\text{proportion of red peanut candy})$$
$$= -.0234$$

Hence

$$(\text{proportion of red plain candy}) =$$
$$(\text{proportion of red peanut candy}) - .0234$$

We "predict" that the proportion of red plain candy is lower than the proportion of red peanut candy, but the 95% confidence interval allows that it might be as much as 3.91% larger (or as much as 8.59% smaller).

**Technology Example 52.3 (TI-84).** Repeat example 52.2 using a TI-84.

1. Open the **2-PropZInt** menu with ⟨stat⟩ ⟩⟩ TESTS ⟩⟩ 2-PropZInt ⟨enter⟩

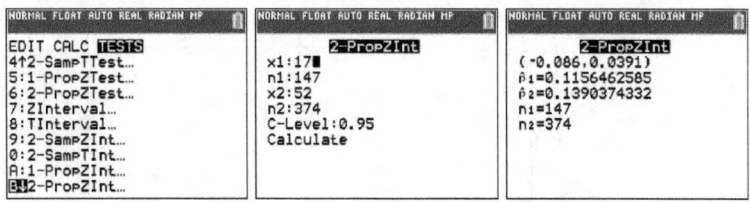

a) For **x1** enter **17** (the number of reds in the first bag).
b) For **n1** enter **147** (the size of the first bag).
c) For **x2** enter **52** (the number of reds in the second bag).

d) For **n2** enter **374** (the size of the second bag).
e) For **C-Level** enter **0.95** (for a 95% confidence level).
f) Navigate to **Calculate** and press enter.

2. The results appear immediately on the **2-PropZInt** output screen. The confidence interval is shown at the top as $(-086, 0.0391)$, consistent with the results of example 52.2. The calculated values of $\hat{p}_1 = 0.1156463585$, $\hat{p}_2 = 0.1390374332$ are also shown.

**Technology Example 52.4 (R).** Repeat example 52.2 in R.

We can solve this problem with the function **prop.test()**.

```
> reds=c(17,52)
> bags=c(147,374)
> prop.test(reds,bags,correct=FALSE)

        2-sample test for equality of
        proportions without continuity
        correction

data:  reds out of bags
X-squared = 0.50251, df = 1,
p-value = 0.4784
alternative hypothesis: two.sided
95 percent confidence interval:
 -0.08585836  0.03907601
sample estimates:
   prop 1    prop 2
0.1156463 0.1390374
```

The option **correct=FALSE** turns off a correction called the Yate's continuity correction, which we do not discuss in this book. The resulting confidence interval of $-.085836$ and $0.03907601$ is consistent with the results of example 52.2.

# Exercises

1. Try the example with the m&m's on your own. Buy bags of two different kind: plain, dark chocolate, peanut butter, mint, caramel, peanut, almond, pretzel, etc. Treating each bag as a random sample, count the numbers of your favorite color in each bag. Then find a confidence interval on the difference in proportions.

2. You are trying to decide which class to take. Professor Smith passed 19 of 24 students last semester, while Professor Jones passed 18 of 21 students. Find a 98% confidence interval on the difference in proportion of students who passed classes with the two instructors

3. A friend told you that when

FAQ 52.  Difference of Proportions, Confidence Interval

you flip coins, American quarter-dollars are more likely to land on heads then American half-dollar coins. You do an experiment and flip one of each coin (a quarter and a half dollar) fifteen times. The quarter gets 8 heads and the half dollar gets 7 heads. Calculate a 95% confidence interval on the difference of proportions.

4. Two weeks before the national elections a survey of 1043 individuals who plan to vote indicates that 490 will vote for John Adams and the rest plan to vote for George Washington. One week later, the survey is repeated with 725 respondents, 365 who say they will vote for Adams. Find a 95% confidence interval on the difference of proportions between the first survey and the second survey. Do you think the difference in survey results might indicate a trend in voter preferences? Could Washington loose the election?

# FAQ 53. How Do I Do a Hypothesis Test on the Difference Between Two Proportions?

Suppose we take a sample of size $n_1$ from population 1 and a sample of size $n_2$ from population $n_2$. In order to perform a hypothesis test against the null hypothesis $H_0 : p_1 - p_2 = 0$ (or equivalently, against the null hypothesis $H_0 : p_1 = p_2$), where $p_1$ and $p_2$ are sample proportions, we need to define a new quantity called the **pooled proportion**. The pooled proportion is a kind of average of the two sample proportions, but it puts more weight on the sample that has more data. The **sample proportions** are defined as

$$\hat{p}_1 = \frac{(\text{number of successes in sample 1})}{n_1}$$

$$\hat{p}_2 = \frac{(\text{number of successes in sample 2})}{n_2}$$

The **pooled sample proportion** $\hat{p}$ is obtained by combining the success and failures from both samples together:

$$\hat{p} = \frac{\text{total number of successes in both samples put together}}{\text{total count of of both samples}}$$
$$= \frac{(\text{successes in sample 1}) + (\text{successes in sample 2})}{n_1 + n_2}$$

**Example 53.1.** Suppose you are comparing the proportion of green candy distributed in bags of two different types of holiday m&m's. In a sample bag of 157 peanut holiday m&m's there are 72 greens, while in a bag of 270 dark chocolate holiday m&m's there are 149 greens. We assign the peanut to population 1, and the dark chocolate to population 2. Then $n_1 = 157$ and $n_2 = 270$. The number of successes in sample 1 is 72 and the number of successes in sample 2 is 149. The sample proportions are

$$\hat{p}_1 = \frac{72}{157} = 0.4586$$

$$\hat{p}_2 = \frac{140}{270} = 0.5185$$

The pooled sample proportion is

$$\hat{p} = \frac{72 + 140}{157 + 270} = \frac{212}{427} = 0.4965$$

---

## Pooled Sample Proportion

Let $\hat{p}_1$ and $\hat{p}_2$ be the sample proportions calculated for some variable as measured in two samples of size $n_1$ and $n_2$. Then the pooled proportion is

$$\hat{p} = w_1 \hat{p}_1 + w_2 \hat{p}_2$$

where $w_1 = n_1/(n_1 + n_2)$ and $w_2 = n_2/(n_1 + n_2)$.

---

The pooled proportion is really a **weighted sample proportion**. Each sample is weighted by the proportion of data that is in each sample:

$$(\text{weight})_1 = \frac{n_1}{n_1 + n_2}$$

$$(\text{weight})_2 = \frac{n_2}{n_1 + n_2}$$

Then

$$\hat{p} = (\text{weight})_1 \times \hat{p}_1 + (\text{weight})_2 \times \hat{p}_2$$

(**Continuation of example 53.1.**) Repeating the calculation of the pooled sample proportion as a weighted sample proportion, we find that

$$(\text{weight})_1 = \frac{157}{157 + 270} = 0.3677$$

---

FAQ 53.  Difference of Proportions Hypotheses Test

$$(\text{weight})_2 = \frac{270}{157 + 270} = 0.6323$$

and therefore the pooled or weighted sample proportion is

$$\hat{p} = (\text{weight})_1 \times \hat{p}_1 + (\text{weight})_2 \times \hat{p}_2$$
$$= (0.3677)(0.4586) + (0.6323)(0.5185)$$
$$= 0.4965$$

---

### Hypothesis Test for Difference of Proportions

To test against the hypothesis $H_0 : p_1 = p_2$ use the statistic

$$z = \frac{\hat{p}_1 - \hat{p}_2}{\sqrt{\hat{p}(1 - \hat{p})\left(\dfrac{1}{n_1} + \dfrac{1}{n_2}\right)}}$$

on simple random samples of size $n_1$ and $n_2$ from normally distributed populations. Here $\hat{p}_1$ and $\hat{p}_2$ are the sample proportions and $\hat{p}$ is the pooled sample proportion. The statistic $z$ has a standard normal distribution $(N(0, 1))$ for sufficiently large sample sizes.

---

The procedure for performing a hypothesis test against the null hypothesis $H_0 : p_1 - p_2 = 0$ follows.

1. **State the problem** you are trying to solve in words.

   **Example 53.2.** A friend has told you that she read the company that makes Skittles has changed the proportion of Skittles that are blue, but she can't remember by how much. Being of a naturally inquisitive nature, you decide to find out. Fortunately, you are a pack rat and have a five year old bag of unopened Skittles in the back of your cupboard. You will use this as one random sample, and a new bag as the other. You pose the following question: "has the proportion of blue Skittles changed?"

2. **Formulate the alternate hypothesis** that is appropriate to the problem. There are three possibilities:

   - $H_A : p_1 > p_2$ (this is a one-sided hypothesis)
   - $H_A : p_1 < p_2$ (this is a one-sided hypothesis)

Figure 53.1.: Outline of the steps in performing a hypothesis test for the difference of two proportions.

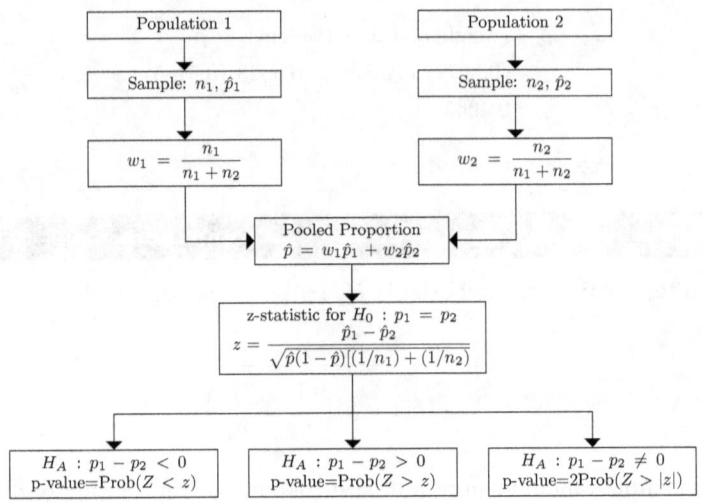

- $H_A: p_1 \neq p_2$ (this is a two-sided hypothesis)

(**Continuation of example 53.2.**) Since you think the proportion might have changed, but you have no idea which way, you will perform a two-sided test $H_A: p_1 \neq p_2$.

3. **State $\alpha$, the desired significance level.**

(**Continuation of example 53.2.**) You decide to perform the test at a 5% significance level, i.e., $\alpha = 0.05$.

4. **Obtain independent random samples** of size $n_1$ and $n_2$ from distinct, normally distributed populations.

(**Continuation of example 53.2.**) The older bag has $n_1 = 327$ skittles in it. The newer bag only has $n_2 = 296$. Same great test, less filling.

5. **Count the number of successes** for the desired variable and determine the sample proportions $\hat{p}_1$ and $\hat{p}_2$.

(**Continuation of example 53.2.**) The old bag has 71 blue skittles. The new bag has 52 blues. The corresponding sample proportions are $\hat{p}_1 = 71/327 = 0.217$ and $\hat{p}_2 = 52/296 = 0.176$.

6. Determine the **pooled proportion**

$$\hat{p} = \frac{n_1\hat{p}_1}{n_1 + n_2} + \frac{n_2\hat{p}_2}{n_1 + n_2}$$

(**Continuation of example 53.2.**) The pooled proportion is

$$\hat{p} = \frac{(327)(0.217)}{327 + 296} + \frac{(296)(0.176)}{327 + 296} = 0.197$$

7. Calculate the **two-proportion pooled z-statistic:**

$$z = \frac{\hat{p}_1 - \hat{p}_2}{\sqrt{\hat{p}(1 - \hat{p})\left(\dfrac{1}{n_1} + \dfrac{1}{n_2}\right)}}$$

(**Continuation of example 53.2.**)

$$z = \frac{0.217 - 0.176}{\sqrt{0.197(1 - 0.197)\left(\dfrac{1}{327} + \dfrac{1}{296}\right)}} = 1.30$$

8. **Determine the p-value**, e.g., by looking up the appropriate tail probability in a table of the normal distribution such as table A.1.

(**Continuation of example 53.2.**) This is a two-sided test so the appropriate p-value is

$$\begin{aligned}
\text{p-value} &= 2 \times \text{Prob}(z > 1.30) \\
&= 2(1 - \text{TableArea}(1.30)) \\
&= 2(1 - .9032) = 0.1936
\end{aligned}$$

where $\text{TableArea}(z)$ is the look-up value in a table of the standard normal distribution, such as in table A.1.

9. Make a conclusion by **comparing the p-value with** $\alpha$.

(**Continuation of example 53.2.**) The p-value is over 19%, which exceeds the chosen $\alpha$ value of 5%. The observed variation or worse will be seen as often as about one time in five when $H_0$ is true[1], so it is quite likely

---

[1]Because exactly one time in five is 20% which is close to 19%, or, 1/0.1936=5.165.

that this difference could have occurred due to normal random variation. We conclude that the observations are not significant at the five percent level and do not provide any evidence that would support rejecting the null hypothesis. There is insufficient evidence to support the claim that the proportion of blue Skittles has changed.

**Technology Example 53.3 (TI-84).** Repeat example 53.2 using a TI-84.

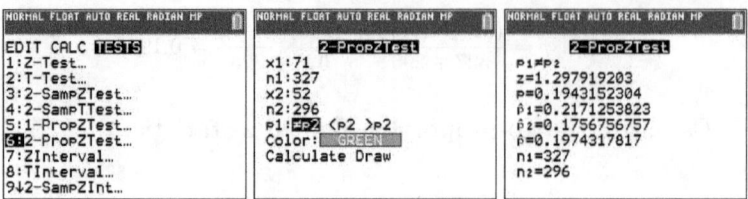

1. Open the **2-PropZTest** menu with [stat] ⟩ TESTS ⟩ [2-PropZTest] [enter].
   a) For **x1** enter **71** (the number of success (blues) in the first bag).
   b) For **n1** enter **327** (the (sample) size of the first bag).
   c) For **x2** enter **52** (the number of success (blues) in the first bag).
   d) For **n2** enter **296** (the (sample) size of the second bag).
   e) Select **p1≠p2** to test the alternate hypothesis $H_A : p_1 \neq p_2$.
   f) Navigate to **Calculate** and press [enter].
2. The solution appears immediately as the **2-PropZTest** output screen. The value of the z statistic is shown as $z = 1.297919203$; the p value is $p = 0.1943152304$; the individual proportions are $\hat{p}_1 = 0.2171253823$ and $\hat{p}_2 = 0.175675675$; and the pooled proportion is $\hat{p} = 0.19743178174$, consistent with example 53.2.

**Technology Example 53.4 (R).** Repeat example 53.2 in R.

We can solve this with **prop.test()** in R.

```
> blues=c(71,52)
> bagsizes=c(327,296)
> prop.test(blues,bagsizes,correct=FALSE)

        2-sample test for equality of proportions
        without continuity correction

data:  blues out of bagsizes
X-squared = 1.6846, df = 1, p-value = 0.1943
alternative hypothesis: two.sided
95 percent confidence interval:
```

FAQ 53.  Difference of Proportions Hypotheses Test

```
-0.02080987   0.10370928
sample estimates:
   prop 1      prop 2
0.2171254 0.1756757
```

As before, the predicted p-value is 0.1943 so there is not sufficient evidence to accept the alternative or reject the null hypothesis.

## Exercises

1. In a bag of 174 m&m candies you count out 43 reds and 37 blues. Is this evidence that there is a difference in proportion of the different colors? Assume the bag is a simple random sample of all candy sold.

2. In a poll of 2352 Florida voters the Saturday before the 2016 US Presi-dential election, 48% of the respondents indicated that they would vote for Trump and 45% would vote for Clinton. On the same day a national of 2700 voters indicated that 47% voters nationally would support Clinton and 43% would support Trump. Is this evidence of a difference in support between voters in Florida and voters nationally?

# FAQ 54. How Do I Find The Marginal Distribution in a Two Way Table?

In a **two way table** information is displayed across two different variables.

**Example 54.1.** Suppose a student has a choice of taking the same introductory statistics class at either 9:00 AM, 2:00 PM, or 7:00 PM. To determine which class to take, the student looks at an online grade aggregator, and finds the following grade information from last semester.

|                      | 9:00 AM Class | 2:00 PM Class | 7:00 PM Class |
|----------------------|:-------------:|:-------------:|:-------------:|
| Number of A's Given  | 1             | 1             | 9             |
| Number of B's Given  | 8             | 6             | 16            |
| Number of C's Given  | 27            | 14            | 10            |
| Number of D's Given  | 3             | 18            | 7             |
| Number of F's Given  | 1             | 16            | 4             |

Disregarding the possibilities that this grade variation could be due to other factors (besides instructor) such as time of day, student population, etc., a student may decide to enroll by choosing the instructor who teaches the course with the highest likelihood of passing.

Two way tables like this give us two types of information: **marginal distributions** and **conditional distributions**.

The **Marginal Distributions** for this data allows us to look at the data across both the rows and the columns.

(**Continuation of example 54.1.**) The marginal distributions can answer the following questions:

- What is the likelihood of passing, regardless of which class is taken?
- What is the likelihood of a student choosing to take this class at a particular time of day, e.g., morning, afternoon, or night?

The **Conditional Distribution** exams the results of one variable assuming the other variable is fixed to a particular value.

(**Continuation of example 54.1.**) The conditional distributions may answer the following questions for this data:

- What is the likelihood of passing the class given that the student took the class at a particular time of day?
- What is the likelihood that the student received a C, given that the student did not take the class in the morning?

To find the marginal distributions, we add an extra row and column. These are called the **margins** of the table.

1. **Add one extra row and column to the table.** These are called the **margins**. Label the row and column headers **sum**.

   (**Continuation of example 54.1.**) The extra row and column are shaded.

   | | 9:00 AM Class | 2:00 PM Class | 7:00 PM Class | Sum |
   |---|---|---|---|---|
   | Number of A's Given | 1 | 1 | 9 | |
   | Number of B's Given | 8 | 6 | 16 | |
   | Number of C's Given | 27 | 14 | 10 | |
   | Number of D's Given | 3 | 18 | 7 | |
   | Number of F's Given | 1 | 16 | 4 | |
   | Sum | | | | |

2. **Total each column** add put the sum in the row on the bottom.

**Total each row** and put the sum on the column in the right. Leave the cell at the bottom right blank.

(**Continuation of example 54.1.**) In the example, the numbers at the bottom will give the total number of students in each class. The numbers in the row on the right will give the total number of students who got a A, a B, etc.

|  | 9:00 AM Class | 2:00 PM Class | 7:00 PM Class | Sum |
|---|---|---|---|---|
| Number of A's Given | 1 | 1 | 9 | 11 |
| Number of B's Given | 8 | 6 | 16 | 30 |
| Number of C's Given | 27 | 14 | 10 | 51 |
| Number of D's Given | 3 | 18 | 7 | 28 |
| Number of F's Given | 1 | 16 | 4 | 21 |
| Sum | 40 | 55 | 46 | |

3. **Add up** either all the numbers in the bottom row, or all the numbers in the right column. These totals should be the same, unless you made a mistake. Put this number in the bottom corner. This is your total sample size.

(**Continuation of example 54.1.**)

|  | 9:00 AM Class | 2:00 PM Class | 7:00 PM Class | Sum |
|---|---|---|---|---|
| Number of A's Given | 1 | 1 | 9 | 11 |
| Number of B's Given | 8 | 6 | 16 | 30 |
| Number of C's Given | 27 | 14 | 10 | 51 |
| Number of D's Given | 3 | 18 | 7 | 28 |
| Number of F's Given | 1 | 16 | 4 | 21 |
| Sum | 40 | 55 | 46 | 141 |

4. **To find the distribution**, add one more row to the bottom, and one more column to the right. Label these **proportion**.

(**Continuation of example 54.1.**)

|  | 9:00 AM Class | 2:00 PM Class | 7:00 PM Class | Sum | Row Proportion |
|---|---|---|---|---|---|
| Number of A's Given | 1 | 1 | 9 | 11 | |
| Number of B's Given | 8 | 6 | 16 | 30 | |
| Number of C's Given | 27 | 14 | 10 | 51 | |
| Number of D's Given | 3 | 18 | 7 | 28 | |
| Number of F's Given | 1 | 16 | 4 | 21 | |
| Sum | 40 | 55 | 46 | 141 | |
| Column Proportion | | | | | |

5. **Determine the row proportions**: Fill in the each entry in column on the right with the corresponding row sum divided by the table total.

$$\text{row proportion} = \frac{\text{row total}}{\text{table total}}$$

(**Continuation of example 54.1.**)

| | 9:00 AM Class | 2:00 PM Class | 7:00 PM Class | Sum | Row Proportion |
|---|---|---|---|---|---|
| Number of A's Given | 1 | 1 | 9 | 11 | 11/141=0.078 |
| Number of B's Given | 8 | 6 | 16 | 30 | 30/141=0.213 |
| Number of C's Given | 27 | 14 | 10 | 51 | 51/141=0.362 |
| Number of D's Given | 3 | 18 | 7 | 28 | 28/141=0.199 |
| Number of F's Given | 1 | 16 | 4 | 21 | 21/141=0.149 |
| Sum | 40 | 55 | 46 | 141 | |
| Column Proportion | | | | | |

6. **Determine the column proportions**: fill in each entry in the row on the bottom with the corresponding column sum divided by the table total.

$$\text{colum proportion} = \frac{\text{column total}}{\text{table total}}$$

(**Continuation of example 54.1.**)

| | 9:00 AM Class | 2:00 PM Class | 7:00 PM Class | Sum | Row Proportion |
|---|---|---|---|---|---|
| Number of A's Given | 1 | 1 | 9 | 11 | 0.078 |
| Number of B's Given | 8 | 6 | 16 | 30 | 0.213 |
| Number of C's Given | 27 | 14 | 10 | 51 | 0.362 |
| Number of D's Given | 3 | 18 | 7 | 28 | 0.199 |
| Number of F's Given | 1 | 16 | 4 | 21 | 0.149 |
| Sum | 40 | 55 | 46 | 141 | |
| Column Proportion | 40/141=0.284 | 55/141=0.390 | 46/141=0.326 | | |

7. To find the marginal distribution in percent rather than proportion, multiply by 100.
(**Continuation of example 54.1.**)

| | 9:00 AM Class | 2:00 PM Class | 7:00 PM Class | Sum | Proportion (or percent) |
|---|---|---|---|---|---|
| Number of A's Given | 1 | 1 | 9 | 11 | 0.078 (7.8%) |
| Number of B's Given | 8 | 6 | 16 | 30 | 0.213 (21.3%) |
| Number of C's Given | 27 | 14 | 10 | 51 | 0.362 (36.2%) |
| Number of D's Given | 3 | 18 | 7 | 28 | 0.199 (19.9%) |
| Number of F's Given | 1 | 16 | 4 | 21 | 0.149 (14.9%) |
| Sum | 40 | 55 | 46 | 141 | |
| Proportion (or %) | 0.284 (28.4%) | 0.390 (39%) | 0.326 (32.6%) | | |

FAQ 54. Marginal Distribution

8. To check your work, verify that your column percentages add up to 100%, and that your row percentages also add up to 100%.

9. You have found two marginal distributions in this way.

   a) **The Row Distribution.**
      (**Continuation of example 54.1.**)
      Distribution of Grades across all classes.

|  | Number | Proportion | Percent |
|---|---|---|---|
| Number of A's Given | 11 | 0.078 | 7.8% |
| Number of B's Given | 30 | 0.213 | 21.3% |
| Number of C's Given | 51 | 0.362 | 36.2% |
| Number of D's Given | 28 | 0.199 | 19.9% |
| Number of F's Given | 121 | 0.149 | 14.9% |
| Total Number of Grades | 141 | 1.0 | 100% |

   b) **The Column distribution.**
      (**Continuation of example 54.1.**)
      Distribution of Students across all classes.

|  | Number | Proportion | Percent |
|---|---|---|---|
| Students in 9:00 AM Class | 40 | 0.284 | 28.4% |
| Students in 2:00 PM Class | 55 | 0.390 | 39.0% |
| Students in 7:00 PM Class | 46 | 0.326 | 32.6% |
| Total Number of Students | 141 | 1.0 | 100% |

**Technology Example 54.2 (R).** Find the row and column sums of the table in example 54.1 using R.

We start by creating a data frame with data in the table. We can define each column as a vector of grades, containing the number of A's, B's, C's, D's and F's in the corresponding class; there are three classes. We create the frame with **data.frame** and label the rows with **row.names**.

```
>morning=c(1,8,27,3,1)
> afternoon=c(1,6,14,18,16)
> evening=c(9,16,10,7,4)
> classes=data.frame(morning,afternoon,evening)
> row.names(classes)=c("A","B","C","D","F")
> classes
  morning afternoon evening
A       1         1       9
B       8         6      16
C      27        14      10
D       3        18       7
F       1        16       4
```

We can then get the row and column sums with **rowSums** and **colSums**.

```
> rowSums(classes)
 A  B  C  D  F
11 30 51 28 21
> colSums(classes)
  morning afternoon   evening
       40        55        46
```

**Technology Example 54.3 (Spreadsheet).** Find the row and column sums of the table in example 54.1 using a spreadsheet.

1. Begin by writing the data into a spreadsheet. The row and column labels are optional, but you can put them there if you like.
2. Use the **sum** function to collect the sums of each row. Do not type in the row and column locations, use the mouse to collect the data locations.
3. You only have to create the row sum once; you can cut and paste it to the other rows and the formulas will automatically update themselves.

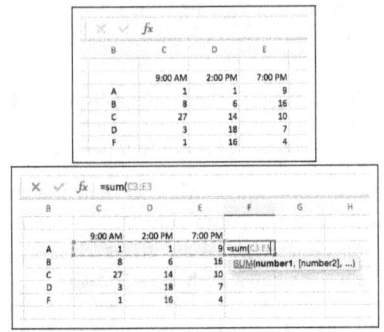

4. Repeat the last two steps for the column sums.

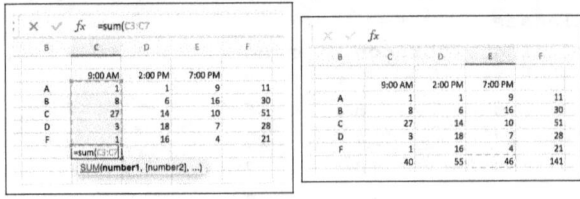

5. To toggle between showing the results in the cells and the formulas in the cells, click on Formulas ⟩⟩ Show Formulas.

FAQ 54. Marginal Distribution

| | B | C | D | E | F |
|---|---|---|---|---|---|
| | | 9 AM | 2 PM | 7 PM | |
| A | | 1 | 1 | 9 | =SUM(C3:E3) |
| B | | 8 | 6 | 16 | =SUM(C4:E4) |
| C | | 27 | 14 | 10 | =SUM(C5:E5) |
| D | | 3 | 18 | 7 | =SUM(C6:E6) |
| F | | 1 | 16 | 4 | =SUM(C7:E7) |
| | | =SUM(C3:C7) | =SUM(D3:D7) | =SUM(E3:E7) | =SUM(F3:F7) |

In a conditional distribution, we form a new table from the raw data by throwing away everything that does not meet the condition. Then we repeat the process with the remaining table. We can condition either on a row variable or a column variable.

1. Formulate the question in words.
   **Example 54.4.** We begin with the results of example 54.1.

| | 9:00 AM Class | 2:00 PM Class | 7:00 PM Class | Sum |
|---|---|---|---|---|
| Number of A's Given | 1 | 1 | 9 | 11 |
| Number of B's Given | 8 | 6 | 16 | 30 |
| Number of C's Given | 27 | 14 | 10 | 51 |
| Number of D's Given | 3 | 18 | 7 | 28 |
| Number of F's Given | 1 | 16 | 4 | 21 |
| Sum | 40 | 55 | 46 | 141 |

   We will ask two questions about this table.

   a) In example 54.4A we ask for the probability of getting an A if you only take daytime classes.

   b) in example 54.4B we ask how the chances of passing (not getting an F) in the morning class compare with the non-morning classes?

2. Identify the appropriate column and row variables.
   **(Continuation of example 54.4A.)**
   Daytime classes are given at 9:00 AM and 2:00 PM. These are described by the column variable.

   **(Continuation of example 54.4B.)**
   Passing a class means not getting an F. This is described by the row variable. Being in a morning class is described by the time day, which is given by a column variable.

3. Remove or aggregate rows or columns as required.
   **(Continuation of example 54.4A.)**
   To get a total of the daytime classes, we form a mini-table that only contains the daytime data.

---

|  | 9:00 AM Class | 2:00 PM Class | Row Sum |
|---|---|---|---|
| Number of A's Given | 1 | 1 | |
| Number of B's Given | 8 | 6 | |
| Number of C's Given | 27 | 14 | |
| Number of D's Given | 3 | 18 | |
| Number of F's Given | 1 | 16 | |
| Sum | 40 | 95 | |

(**Continuation of example 54.4B.**)
Passing a class means not getting an F. We form a new mini-table that aggregates (adds up) all the passing-grades into a single row.

|  | 9:00 AM Class | 2:00 PM Class | 7:00 PM Class | Row Sum |
|---|---|---|---|---|
| Passed | 1+8+27+3=39 | 1+6+14+18=39 | 9+16+10+7=42 | |
| Failed | 1 | 16 | 4 | |
| Sum | 40 | 55 | 46 | |

4. Calculate the appropriate marginal distribution of the modified table (after you have removed rows or columns) or desired statistic.
(**Continuation of example 54.4A.**)
Calculate the row sum of the mini-table.

|  | 9:00 AM Class | 2:00 PM Class | Row Sum |
|---|---|---|---|
| Number of A's Given | 1 | 1 | 2 |
| Number of B's Given | 8 | 6 | 14 |
| Number of C's Given | 27 | 14 | 41 |
| Number of D's Given | 3 | 18 | 21 |
| Number of F's Given | 1 | 16 | 17 |
| Sum | 40 | 55 | 95 |

(**Continuation of example 54.4B.**)
Add up the row sums.

|  | 9:00 AM Class | 2:00 PM Class | 7:00 PM Class | Row Sum |
|---|---|---|---|---|
| Passed | 39 | 39 | 42 | 120 |
| Failed | 1 | 16 | 4 | 21 |
| Sum | 40 | 55 | 46 | 141 |

At this point we notice that only one class is offered in the morning, so we want to aggregate the non-morning classes as well. We form a new mini-table. The row sums do not change but the entries in the table are aggregated.

|  | Morning Class | Afternoon and Evening | Sum |
|---|---|---|---|
| Passed | 39 | 39+42=81 | 120 |
| Failed | 1 | 16+4=20 | 21 |
| Sum | 40 | 101 | 141 |

5. Find the desired conditional distribution and, if requested, any desired descriptive statistics.

**(Continuation of example 54.4A.)**
Since there are only twp A's the chances of getting an A are 2/95 or 2.1%.

The corresponding conditional distribution is the distribution of grades in morning classes:

| | As a Number | As a Proportion | As a Percent |
|---|---|---|---|
| Number of A's Given | 2 | 2/95=0.021 | 2.1% |
| Number of B's Given | 14 | 4/95 = 0.042 | 4.2% |
| Number of C's Given | 41 | 41/95 = 0.432 | 43.2% |
| Number of D's Given | 21 | 21/95 = 0.221 | 22.1% |
| Number of F's Given | 17 | 17/95 = 0.179 | 17.9% |
| Sum | 95 | 1.0 | 100% |

**(Continuation of example 54.4B.)**
The final aggregated table looked like this:

| | Morning Class | Afternoon and Evening | Sum |
|---|---|---|---|
| Passed | 39 | 81 | 120 |
| Failed | 1 | 20 | 21 |
| Sum | 40 | 101 | 141 |

The probability of passing a morning class were

$$P(\text{Pass AM Class}) = \frac{\text{Number of Students in AM Class That Passed}}{\text{Total Number of Students in AM Class}}$$

$$= \frac{39}{40} = .975 \text{ or } 97.5\%$$

The probability of passing, given that you took a non-morning class, were

$$P(\text{Pass PM Class}) = \frac{\text{Number of Students in PM Class That Passed}}{\text{Total Number of Students in PM Class}}$$

$$= \frac{81}{101} = 0.802 \text{ or } 80.2\%$$

Thus based on this sample, the students who took the morning class stood a better chance of passing.

# Exercises

1. A study group on campus interviews 740 students on campus and asks them how many hours per week they spent studying per week. Students identified themselves by major in one of several groups, and as class year. Results are given in the table 54.1.

   a) Find the marginal distributions.
   b) Which major spends the most time studying?
   c) Which class year spends the least time studying?

Table 54.1.: Data for exercise 1.

| Class | Freshman | Sophomore | Junior | Senior |
|---|---|---|---|---|
| Engineering | 14 | 15 | 16.5 | 19.3 |
| Humanities | 15 | 15.5 | 16 | 17 |
| Business | 12 | 13 | 14 | 14.5 |
| Sciences | 14.5 | 15 | 15.5 | 16 |
| Arts and Communication | 15 | 14.5 | 16 | 18 |
| Social Sciences | 13 | 17 | 16 | 17 |
| Human Development | 15 | 15.7 | 16.3 | 17.6 |

Table 54.2.: Data for exercise 2.

| | Red | Blue | Green | Orange | Brown | Yellow |
|---|---|---|---|---|---|---|
| Milk Chocolate M&M | 15 | 10 | 23 | 18 | 12 | 11 |
| Dark Chocolate M&M | 12 | 5 | 14 | 12 | 10 | 11 |
| Peanut M&M | 32 | 0 | 27 | 0 | 0 | 0 |
| Mint M&M | 0 | 0 | 35 | 0 | 0 | 0 |
| Peanut Butter M&M | 15 | 12 | 12 | 11 | 10 | 12 |

2. Several candy dishes are scattered around various tables at a party. Overall, they contain an enormous amount of M&M's. You want to find the distribution so you take one cup of candy from one dish and count them out. You assume that that all of the dishes were well mixed beforehand. The results are shown in table 54.2.

a) Find the marginal distributions.

b) Classify the candy as ei-

ther plain (milk chocolate or dark chocolate) or flavored (peanut, peanut butter, or mint). Find the conditional distribution of green candy as either plain or flavored.

c) Some people have an allergy to peanuts. What is the probability that a randomly selected candy will be either peanut butter or peanut?

d) Find the probability of getting a dark chocolate candy that is neither red nor blue.

# FAQ 55. How Do I Calculate Expected Values in a Two-Way Table?

As we saw in chapter 54, two way tables can be used to to compare the distributions of two categorical variables.

**Example 55.1.** Here are the counts of m&m's by color in a candy dish.

|                | Red | Blue | Green | Orange | Brown | Yellow |
|----------------|-----|------|-------|--------|-------|--------|
| Milk Chocolate | 10  | 3    | 18    | 12     | 11    | 17     |
| Peanut         | 12  | 7    | 17    | 16     | 14    | 13     |
| Peanut Butter  | 44  | 10   | 32    | 11     | 10    | 12     |

We can get two new distributions from the row and column sums:

|                | Red | Blue | Green | Orange | Brown | Yellow | Sum |
|----------------|-----|------|-------|--------|-------|--------|-----|
| Milk Chocolate | 10  | 3    | 18    | 12     | 11    | 17     | 71  |
| Peanut         | 12  | 7    | 17    | 16     | 14    | 13     | 79  |
| Peanut Butter  | 44  | 10   | 32    | 11     | 10    | 12     | 119 |
| Sum            | 66  | 20   | 67    | 29     | 35    | 42     | 269 |

Here are the two new distributions:

- Distribution of type or flavor

| Milk Chocolate | Peanut | Peanut Butter |
|----------------|--------|---------------|
| 71             | 79     | 119           |

- Distribution of color

| Red | Blue | Green | Orange | Brown | Yellow |
|-----|------|-------|--------|-------|--------|
| 66  | 20   | 67    | 29     | 35    | 42     |

If we assume that the two marginal distributions are **independent** of one another, then we can predict the expected value in each entry of a two way table. Two distributions are said to be independent if the probability described by one is not affected by the value in the other.

**Example 55.2.** By assuming that the two marginal distributions in example 55.1 are independent we would be assuming, for example, that the chances of an m&m being red, is the same, whether it is a plain m&m, a peanut m&m, or a peanut butter m&m. Since the candies were mixed together in a dish, it is possible that the host of the party who put them there arranged for equal distributions. However, the company that makes m&m's does not, in general, use the same distribution of colors for each flavor. So if we just mix bags of candy that we bought in the store, this assumption would not be correct.

**Example 55.3.** You call for an Uber to pick you up after a late night doing statistics homework at your local pub. In your neighborhood there are 23 black Toyotas, 11 white Toyotas, 14 black Hondas and 17 white Hondas in the size and style of car that you requested.

|        | White Car | Black Car | Row Sum |
|--------|-----------|-----------|---------|
| Toyota | 11        | 23        | 34      |
| Honda  | 17        | 14        | 31      |
| Sum    | 28        | 37        | 65      |

The marginal distributions are independent.

---

### Expected Value in a Two Way Table

If the two variables in a two-way table are **independent random variables** (i.e., they have no relationship to one another), then the expected value in any table location is

$$\text{Expected Value} = \frac{(\text{row total}) \times (\text{column total})}{\text{table total}}$$

where the (row total) and (column total) are the sums in the marginal distributions for the row and column that intersect in that location, and the (table total) is the total value of all the entries in the table.

---

To calculate the expected values in a two way table you must know all the marginal totals and the total for the table.

1. Draw a full two-way table for the data, but leave the spaces where the data values go blank. Fill in ONLY the marginal sums.
   **Example 55.4.** Using the completed two-way table from example 55.1, we have the following:

|                | Red | Blue | Green | Orange | Brown | Yellow | Sum |
|----------------|-----|------|-------|--------|-------|--------|-----|
| Milk Chocolate |     |      |       |        |       |        | 71  |
| Peanut         |     |      |       |        |       |        | 79  |
| Peanut Butter  |     |      |       |        |       |        | 119 |
| Sum            | 66  | 20   | 67    | 29     | 35    | 42     | 269 |

2. Fill in the expected counts into each cell:

$$\text{expected counts} = \frac{(\text{row sum}) \times (\text{colum sum})}{(\text{table total})}$$

(**Continuation of example 55.4.**) For example, the number at the intersection of the row "Peanut" and "Orange" has a value of $(79) \times (29)/259$. Fill in the expressions in each box, leaving space to complete your calculation. Here are the calculations:

| | Red | Blue | Green | Orange | Brown | Yellow | Sum |
|---|---|---|---|---|---|---|---|
| Milk | $\dfrac{71 \times 66}{269}$ | $\dfrac{71 \times 20}{269}$ | $\dfrac{71 \times 67}{269}$ | $\dfrac{71 \times 29}{269}$ | $\dfrac{71 \times 35}{269}$ | $\dfrac{71 \times 42}{269}$ | 71 |
| Peanut | $\dfrac{79 \times 66}{269}$ | $\dfrac{79 \times 20}{269}$ | $\dfrac{79 \times 67}{269}$ | $\dfrac{79 \times 29}{269}$ | $\dfrac{79 \times 35}{269}$ | $\dfrac{79 \times 42}{269}$ | 79 |
| Pea. But. | $\dfrac{119 \times 66}{269}$ | $\dfrac{119 \times 20}{269}$ | $\dfrac{119 \times 67}{269}$ | $\dfrac{119 \times 29}{269}$ | $\dfrac{119 \times 35}{269}$ | $\dfrac{119 \times 42}{269}$ | 119 |
| Sum | 66 | 20 | 67 | 29 | 35 | 42 | 269 |

Completing the calculations, here are the results:

| | Red | Blue | Green | Orange | Brown | Yellow | Sum |
|---|---|---|---|---|---|---|---|
| Milk Chocolate | 17.42 | 5.28 | 17.68 | 10.29 | 9.24 | 11.09 | 71 |
| Peanut | 19.38 | 5.87 | 19.68 | 11.45 | 10.28 | 12.33 | 79 |
| Peanut Butter | 29.2 | 8.85 | 29.64 | 17.25 | 15.48 | 18.58 | 119 |
| Sum | 66 | 20 | 67 | 29 | 35 | 42 | 269 |

**Technology Example 55.5 (TI-84).** The expected values are calculated automatically by the TI-84 as part of the chi-squared test. An illustration is given in example 56.4.

# Exercises

1. Find the expected values for the data table in example 55.3.

2. Find the expected values for the data table in exercise 54.1.

3. Find the expected values for the data table in exercise 54.2.

# FAQ 56. What is a Chi Squared Test of Independence?

Suppose we have a two-way table (see, for example, chapter 54). The $\chi^2$ (**Chi-squared**) **test for independence** gives a way of evaluating the following hypotheses:

$H_0$: The data in the rows and columns are independent

$H_A$: The data in the rows and columns are not independent

To perform this test we calculated the chi-squared ($\chi^2$, where $\chi$ is a letter in the Greek alphabet). The value of $\chi^2$ is actually a measurement of how far away the data is from what we would expect to get if the two marginal distributions were independent, i.e., the rows and columns do not depend on one another.

---

### $\chi^2$ Test Statistic

The $\chi^2$ **test statistic** for a two way table is

$$\chi^2 = \sum_{\text{table}} \frac{(\text{observed value - expected value})^2}{(\text{expected value})}$$

where

- The sum is over all data locations in the table;
- The "observed values" are the raw data values;
- The "expected values" are the expected values (ch. 55).

---

Once we have the value of $\chi^2$, we determine the probability that such a value can occur, assuming that $H_0$ is true, by looking up its value in a chi square table (such as table A.3). In general, **larger values** of $\chi^2$ will be evidence against $H_0$. The p-value for the test is the area under the curve of a chi-squared distribution to the right of a critical value $\chi^*$ that is determined by $\chi^2$ (figure 56.1).

---

Figure 56.1.: Typical $\chi^2$ distribution. The tail probability for a given critical value $\chi^2$ is hatched.

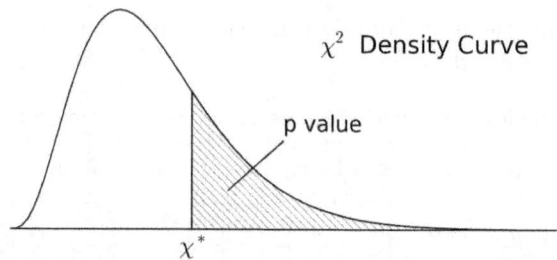

degrees of freedom, where "rows" and "columns" are the numbers of rows and columns in the two-way table.

---

**P-Value for a $\chi^2$-test**

The p-value for a $\chi^2$ test is the probability that that the observed value of $\chi^2$ will occur in a $\chi^2$ distribution with

$$df = (\text{rows-1}) \times (\text{columns-1})$$

degrees of freedom, where "rows" and "columns" are the numbers of rows and columns in the two-way table.

---

The p value decreases rapidly with increasing values of the $\chi^2$ statistic but it increases as more degrees of freedom are added (fig.56.2).

Figure 56.2.: Larger values of $\chi^2$ have smaller p-values; each curve illustrates the relationship for a different number of degrees of freedom.

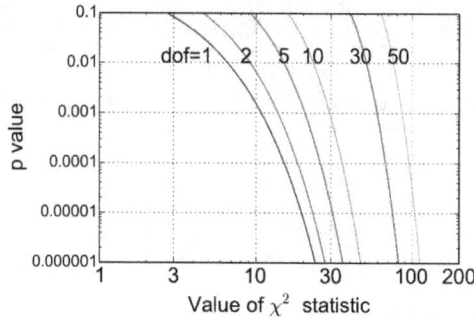

Here is a procedure to test to see if the rows and columns variables are distributed independently.

1. Calculate the **marginal distributions** (row and column sums) and **table total** (ch. 54).

   **Example 56.1.** In example 55.1 we determined the expected values for the colors of m&m's in a bowl of candy. We now ask if the **distributions by color** and the **distributions by type** are independent or not.

   |  | Red | Blue | Green | Orange | Brown | Yellow | Sum |
   |---|---|---|---|---|---|---|---|
   | Milk Chocolate | 10 | 3 | 18 | 12 | 11 | 17 | 71 |
   | Peanut | 12 | 7 | 17 | 16 | 14 | 13 | 79 |
   | Peanut Butter | 44 | 10 | 32 | 11 | 10 | 12 | 119 |
   | Sum | 66 | 20 | 67 | 29 | 35 | 42 | 269 |

   The row and column sums have already been calculated.

2. Calculate the **expected values** for each cell (ch.55).

   **(Continuation of example 56.1.)**
   From example 55.1:

   |  | Red | Blue | Green | Orange | Brown | Yellow | Sum |
   |---|---|---|---|---|---|---|---|
   | Milk Chocolate | 17.42 | 5.28 | 17.68 | 10.29 | 9.24 | 11.09 | 71 |
   | Peanut | 19.38 | 5.87 | 19.68 | 11.45 | 10.28 | 12.33 | 79 |
   | Peanut Butter | 29.2 | 8.85 | 29.64 | 17.25 | 15.48 | 18.58 | 119 |
   | Sum | 66 | 20 | 67 | 29 | 35 | 42 | 269 |

3. **Copy the table**, excluding the row and column sums, writing each expected data value next to it in parenthesis.

   **(Continuation of example 56.1.)**
   Each cell with $O(E)$ values:

   |  | Red | Blue | Green | Orange | Brown | Yellow |
   |---|---|---|---|---|---|---|
   | Milk Chocolate | 10 (17.42) | 3 (5.28) | 18 (17.68) | 12 (10.29) | 11 (9.24) | 17 (11.09) |
   | Peanut | 12 (19.38) | 7 (5.87) | 17 (19.68) | 16 (11.45) | 14 (10.28) | 13 (12.33) |
   | Peanut Butter | 44 (29.2) | 10 (8.85) | 32 (29.64) | 11 (17.25) | 10 (15.48) | 12 (18.58) |

   This new table will allow you to see the $O$ and $E$ values next to each other, so that they can be used together in calculation.

4. Formulate the sum

$$\chi^2 = \sum \frac{[(\text{obs.}) - (\text{exp.})]^2}{\text{exp.}}.$$

   over the entire table.

**(Continuation of example 56.1.)**

$$\chi^2 = \frac{(10-17.42)^2}{17.42} + \frac{(3-5.28)^2}{5.28} + \frac{(18-17.68)^2}{17.68} + \frac{(12-10.29)^2}{10.29} +$$

$$\frac{(11-9.24)^2}{9.24} + \frac{(17-11.09)^2}{11.09} +$$

$$\frac{(12-19.38)^2}{19.38} + \frac{(7-5.87)^2}{5.87} + \frac{(17-19.68)^2}{19.68} + \frac{(16-11.45)^2}{11.45} +$$

$$\frac{(14-10.28)^2}{10.28} + \frac{(13-12.33)^2}{12.33} +$$

$$\frac{(44-29.2)^2}{29.2} + \frac{(10-8.85)^2}{8.85} + \frac{(32-29.64)^2}{29.64} + \frac{(11-17.25)^2}{17.25} +$$

$$\frac{(10-15.48)^2}{15.48} + \frac{(12-18.58)^2}{18.58}$$

$$= 3.1605 + 0.9845 + 0.0058 + 0.2842 + 03352 + 3.1495 +$$

$$2.8103 + 0.2175 + 0.3650 + 1.8081 + 1.3461 + 0.0364 +$$

$$7.5014 + 0.1494 + 0.1879 + 2.2645 + 1.9399 + 2.3303$$

$$= 28.88$$

5. The **degrees of freedom (df) of the table** are given by

$$df = (\text{rows-1}) \times (\text{columns-1})$$

.

**(Continuation of example 56.1.)**  The table of raw data has 3 rows and 6 columns. Thus

$$df = (\text{rows} - 1) \times (\text{columns} - 1) = (3-1) \times (6-1) = 10.$$

6. **Determine the p-value** based on the calculated $\chi^2$ and degrees of freedom of the table.

   a) (Method 1) Use a calculator function or statistical software.

   **(Continuation of example 56.1.)**  Using statistical software or a calculator (see, e.g., example 56.3) button with $\chi^2 = 28.88$ and df=10 gives a p-value of 0.001303.

   b) (Method 2) Use a table of the chi-squared distribution such as table A.3.

      i. Locate the row in the table corresponding to the degrees of freedom (df) of the problem.

---

ii. Move along the row until you find two adjacent values in the row that bracket your calculated value of $\chi^2$, one value smaller, and one value larger. These are called bounding critical values.

iii. Look at the top row of the table. This gives two values of $p$ that bracket the p-value, one larger, and one smaller.

iv. If all of the values in the row df are smaller than your calculated $\chi^2$, then the p-value is smaller than than rightmost value of $p$ in the top row.

**(Continuation of example 56.1.)**

i. Since we have 10 degrees of freedom, we use row 10 of table A.3.

ii. We move from left to right along this row until we find two values that bracket our $\chi^2 = 28.88$ value These values are 27.11 and 29.59.

iii. Projecting from these two numbers to the first row of table A.3, we see that the p-value is bounded by 0.0025 and 0.001. Thus

$$0.0025 < p - value < 0.001$$

**Technology Example 56.2 (TI-84).** Demonstrate how the p-value calculated in step 6 of example 56.1 can be calculated on a TI-84.

This uses the $\chi^2$**cdf** function on the calculator. This function has three arguments: **lower**; **upper** and **df**. It returns the area under a $\chi^2$ cdf with *df* degrees of freedom, between the lower value and the upper value. The find the p-value, we set **lower** to the value of $\chi^2$ and **upper** to a very large number (that represents infinity).

1. From the **DISTR** menu, select $\chi^2$**cdf**: [2nd] [vars] [DISTR⟩⟩$\chi^2$cdf] [enter]
2. On the $\chi^2$**cdf** menu:

   a) For **lower**, enter **28.88** (the value of $\chi^2$);
   b) For **upper**, enter **99999** (or any extremely large number, because we are finding a right-tail area);
   c) For **df** enter **10** (the number of degrees of freedom);
   d) Scroll to **Paste** and press [enter].

3. The command $\chi^2$**cdf(28.88,99999,10)** will appear on your screen as input. Press [enter].
4. The resulting right-tail probability of **0.0013030095** appears on the screen. This is the p-value.

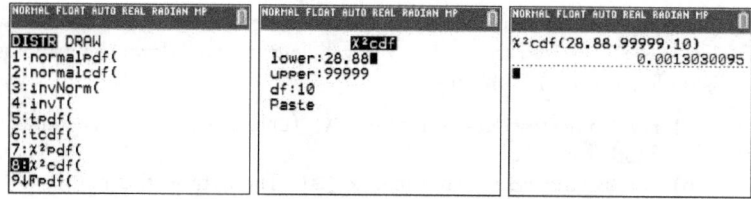

**Technology Example 56.3 (R).** Illustrate how the p-value calculated in step 6 of example 56.1 can be calculated in R.

This can be done with the **pchisq()** function, which returns the cumulative distribution (CDF) of the $\chi^2$ probability function. The full syntax is

**pchisq(chisquared, df, lower.tail=FALSE)**

We have a $\chi^2$ value of $28.88$, and $10$ degrees of freedom. The flag **lower.tail=FALSE** tells the function to return the right tail (area to the right) rather than the area to the left, which is the default.

```
> pchisq(28.88, 10, lower.tail=FALSE)
[1] 0.001303009
```

Thus the p-value is equal to $0.001303009$.

**Technology Example 56.4 (TI-84).** Repeat example 56.1 using a TI-84.

1. Enter the table into a matrix. Open the matrix edit menu with [2nd] [matrix] then select [EDIT]
2. Select the first matrix (named **[A]**) followed by [enter].
3. Set the matrix size to **3 × 6** (the first row of the edit menu), because we have a table with $3$ rows and $6$ columns.
4. Enter the observed data into matrix **[A]** exactly as it is given.

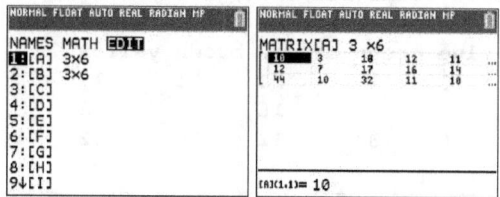

5. When the observed data has been filled in, it is not necessary to enter the expected values. These will be calculated for you automatically.

---

6. From the **TESTS** menu select $\chi^2$**–Test**. To get there, type stat followed by TESTS $\rangle$ $\chi^2$-Test and enter. You will have to scroll down the menu as it is on line **C** (line 13, off screen).

   a) For **Observed** select matrix **[A]** (unless you typed the data into another matrix.);

   b) For **Expected** selected matrix **[B]**. The data in this matrix will be over-written with the expected values. You do not have to use matrix **[B]**; if you prefer, you can write the expected values to another matrix and write the name of the matrix here.

   c) Scroll to **Calculate** and press enter.

7. The result will be displayed on the screen. Here we see that $\chi^2 = 28.88600619$ and the p-value is 0.001300091, consistent with our previous estimate.

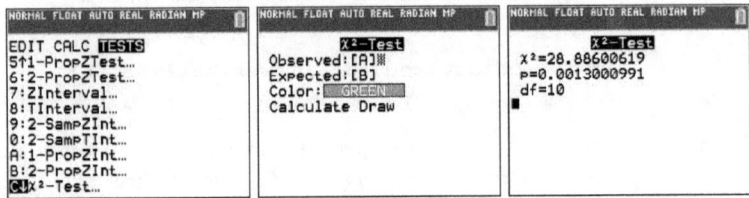

8. If you want to see the expected values, return to the matrix edit screen and select matrix **[B]**.

**Technology Example 56.5 (R).** Repeat example 56.1 in R.

We can do this by putting the observed values into a data frame and using **chisq.test** on the data. First, we put the data into column vectors, and load the columns into a frame:

```
> red=c(10,12,44)
> blue=c(3,7,10)
> green=c(18,17,32)
> orange=c(12,16,11)
> brown=c(11,14,10)
> yellow=c(17,13,12)
> mm=data.frame(red,blue,green,orange,brown,yellow)
> mm
  red blue green orange brown yellow
1  10    3    18     12    11     17
2  12    7    17     16    14     13
3  44   10    32     11    10     12
}
```

Perform the chi squared test with **chisq.test**:

---

```
> chisq.test(mm)

        Pearson's Chi-squared test

data: mm
X-squared = 28.886, df = 10, p-value = 0.0013
```

The calculated p-value is 0.0013, which agrees with out other calculations.

**Technology Example 56.6 (Spreadsheet).** Repeat example 56.1 using a spreadsheet.

1. Enter the observed data into a square block of the table. You may label the rows and columns if you wish, although it is not required.

| F7 | | $f_x$ | | | | | |
|---|---|---|---|---|---|---|---|
| | A | B | C | D | E | F | G |
| 1 | | | | Observed Values | | | |
| 2 | | Red | Blue | Green | Orange | Yellow | Brown |
| 3 | Milk Chocola | 10 | 3 | 18 | 12 | 11 | 17 |
| 4 | Peanut | 12 | 7 | 17 | 16 | 14 | 13 |
| 5 | Peanut Butte | 44 | 10 | 32 | 11 | 10 | 12 |
| 6 | | | | | | | |

2. Compute the row and column sums.

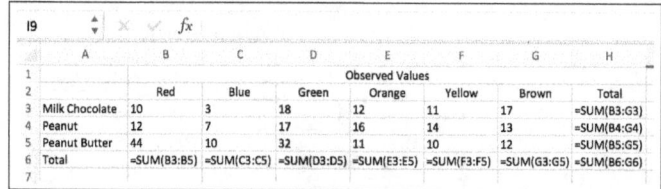

| I9 | | $f_x$ | | | | | | |
|---|---|---|---|---|---|---|---|---|
| | A | B | C | D | E | F | G | H |
| 1 | | | | | Observed Values | | | |
| 2 | | Red | Blue | Green | Orange | Yellow | Brown | Total |
| 3 | Milk Chocolate | 10 | 3 | 18 | 12 | 11 | 17 | =SUM(B3:G3) |
| 4 | Peanut | 12 | 7 | 17 | 16 | 14 | 13 | =SUM(B4:G4) |
| 5 | Peanut Butter | 44 | 10 | 32 | 11 | 10 | 12 | =SUM(B5:G5) |
| 6 | Total | =SUM(B3:B5) | =SUM(C3:C5) | =SUM(D3:D5) | =SUM(E3:E5) | =SUM(F3:F5) | =SUM(G3:G5) | =SUM(B6:G6) |
| 7 | | | | | | | | |

You can toggle the formula view with the by clicking on Formula ⟫ Show Formulas.

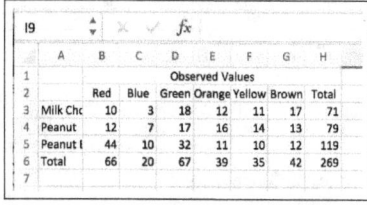

| I9 | | | $f_x$ | | | | | |
|---|---|---|---|---|---|---|---|---|
| | A | B | C | D | E | F | G | H |
| 1 | | | | Observed Values | | | | |
| 2 | | Red | Blue | Green | Orange | Yellow | Brown | Total |
| 3 | Milk Chc | 10 | 3 | 18 | 12 | 11 | 17 | 71 |
| 4 | Peanut | 12 | 7 | 17 | 16 | 14 | 13 | 79 |
| 5 | Peanut l | 44 | 10 | 32 | 11 | 10 | 12 | 119 |
| 6 | Total | 66 | 20 | 67 | 39 | 35 | 42 | 269 |
| 7 | | | | | | | | |

3. Create a second box with the expected values.

| | A | B | C | D | E | F | G | H |
|---|---|---|---|---|---|---|---|---|
| 1 | | | | | Observed Values | | | |
| 2 | | Red | Blue | Green | Orange | Yellow | Brown | Total |
| 3 | Milk Chocolate | 10 | 3 | 18 | 12 | 11 | 17 | =SUM(B3:G3) |
| 4 | Peanut | 12 | 7 | 17 | 16 | 14 | 13 | =SUM(B4:G4) |
| 5 | Peanut Butter | 44 | 10 | 32 | 11 | 10 | 12 | =SUM(B5:G5) |
| 6 | Total | =SUM(B3:B5) | =SUM(C3:C5) | =SUM(D3:D5) | =SUM(E3:E5) | =SUM(F3:F5) | =SUM(G3:G5) | =SUM(B6:G6) |
| 7 | | | | | | | | |
| 8 | | | | | | | | |
| 9 | | | | | Expected Values | | | |
| 10 | Milk Chocolate | =$H3*B$6/$H$6 | =$H3*C$6/$H$6 | =$H3*D$6/$H$6 | =$H3*E$6/$H$6 | =$H3*F$6/$H$6 | =$H3*G$6/$H$6 | =SUM(B10:G10) |
| 11 | Peanut | =$H4*B$6/$H$6 | =$H4*C$6/$H$6 | =$H4*D$6/$H$6 | =$H4*E$6/$H$6 | =$H4*F$6/$H$6 | =$H4*G$6/$H$6 | =SUM(B11:G11) |
| 12 | Peanut Butter | =$H5*B$6/$H$6 | =$H5*C$6/$H$6 | =$H5*D$6/$H$6 | =$H5*E$6/$H$6 | =$H5*F$6/$H$6 | =$H5*G$6/$H$6 | =SUM(B12:G12) |
| 13 | Total | =SUM(B10:B12) | =SUM(C10:C12) | =SUM(D10:D12) | =SUM(E10:E12) | =SUM(F10:F12) | =SUM(G10:G12) | =SUM(H10:H12) |
| 14 | | | | | | | | |

You can toggle the formula view with the button $\boxed{\text{Formula} \gg \text{Show Formulas}}$.

G19    × ✓   fx

| | A | B | C | D | E | F | G | H |
|---|---|---|---|---|---|---|---|---|
| 1 | | | | | Observed Values | | | |
| 2 | | Red | Blue | Green | Orange | Yellow | Brown | Total |
| 3 | Milk Cho | 10 | 3 | 18 | 12 | 11 | 17 | 71 |
| 4 | Peanut | 12 | 7 | 17 | 16 | 14 | 13 | 79 |
| 5 | Peanut I | 44 | 10 | 32 | 11 | 10 | 12 | 119 |
| 6 | Total | 66 | 20 | 67 | 39 | 35 | 42 | 269 |
| 7 | | | | | | | | |
| 8 | | | | | | | | |
| 9 | | | | | Expected Values | | | |
| 10 | Milk Cho | 17.42 | 5.2788 | 17.684 | 10.294 | 9.2379 | 11.086 | 71 |
| 11 | Peanut | 19.383 | 5.8736 | 19.677 | 11.454 | 10.279 | 12.335 | 79 |
| 12 | Peanut I | 29.197 | 8.8476 | 29.639 | 17.253 | 15.483 | 18.58 | 119 |
| 13 | Total | 66 | 20 | 67 | 39 | 35 | 42 | 269 |

4. Find the p-value for a chi-squared test using **CHISQ.TEST(observed, expected)**. The answer appears in the same cell.

I1    × ✓   fx   =CHISQ.TEST(B3:G5,B10:G12)

| | A | B | C | D | E | F | G | H | I |
|---|---|---|---|---|---|---|---|---|---|
| 1 | | | | | Observed Values | | | | 0.0013001 |
| 2 | | Red | Blue | Green | Orange | Yellow | Brown | Total | |
| 3 | Milk Cho | 10 | 3 | 18 | 12 | 11 | 17 | 71 | |
| 4 | Peanut | 12 | 7 | 17 | 16 | 14 | 13 | 79 | |
| 5 | Peanut I | 44 | 10 | 32 | 11 | 10 | 12 | 119 | |
| 6 | Total | 66 | 20 | 67 | 39 | 35 | 42 | 269 | |
| 7 | | | | | | | | | |
| 8 | | | | | | | | | |
| 9 | | | | | Expected Values | | | | |
| 10 | Milk Cho | 17.42 | 5.2788 | 17.684 | 10.294 | 9.2379 | 11.086 | 71 | |
| 11 | Peanut | 19.383 | 5.8736 | 19.677 | 11.454 | 10.279 | 12.335 | 79 | |
| 12 | Peanut I | 29.197 | 8.8476 | 29.639 | 17.253 | 15.483 | 18.58 | 119 | |
| 13 | Total | 66 | 20 | 67 | 39 | 35 | 42 | 269 | |

The calculated p-value is 0.0013.

# Exercises

Perform $\chi^2$ tests for independence on the following data sets, and interpret your results.

1. The following table, which breaks down the number of boys and girls in each of three different classrooms:

| | Room 1 | Room 2 | Room 3 |
|---|---|---|---|
| Boys | 23 | 15 | 13 |
| Girls | 15 | 17 | 17 |

2. The following table, which repre-

sents the inventory for three items
in three different stores :

| | Store A | Store B | Store C |
|---|---|---|---|
| Widgets | 7 | 10 | 14 |
| Wadgets | 6 | 5 | 9 |
| Woodgets | 18 | 12 | 14 |

3. The data from exercise 54.1.

# FAQ 57. What is a Chi Squared Homogeneity Test?

The **chi-squared test for homogeneity** is used to determined if the
distributions of a variable are the same in different populations. We take
a sample from each of two different populations, and then then count the
distributions of the same variable in each sample. We form a $2 \times n$ table,
where $n$ is the number categories in the distribution. We formulate the
following hypotheses

$H_0$: The distributions are the same

$H_A$: The disributions are not the same

**Example 57.1.** Are the color distributions in small bags of m&m's and party-size
bags of m&m's the same?

This can be also be tested the chi-squared technique.

1. **State the null and alternative hypotheses** for the test.

   (**Continuation of example 57.1.**)

   $H_0$: The color distributions are the same

   $H_A$: The color distributions are different

2. **State a confidence level** in terms of an $\alpha$ (alpha) value. Typically $\alpha$ is 0.05 or smaller. Sometimes this is stated in terms of a
   confidence level in percent, where confidence $= 100 \times (1 - \alpha)$, or
   equivalently, $\alpha = (100 - \text{confidence})/100$.

   (**Continuation of example 57.1.**)
   We will use $\alpha = 0.05$ (this is sometimes stated as a 95% confidence level).

3. **Obtain random samples** from each population.

**(Continuation of example 57.1.)**
You purchase one small bag of m&m's, and a large party sized bag of m&m's. You use the entire small bag as one sample, and take a scoop out of the party sized bag for the second sample. This is what you find:

|        | Small Bag | Large Bag |
|--------|-----------|-----------|
| Red    | 10        | 6         |
| Orange | 3         | 10        |
| Yellow | 6         | 5         |
| Green  | 6         | 4         |
| Blue   | 6         | 6         |
| Brown  | 12        | 2         |

4. Calculate the **marginal distributions.**

**(Continuation of example 57.1.)**

|        | Small | Large | Total |
|--------|-------|-------|-------|
| Red    | 10    | 6     | 16    |
| Orange | 3     | 10    | 13    |
| Yellow | 6     | 5     | 11    |
| Green  | 6     | 4     | 10    |
| Blue   | 6     | 6     | 12    |
| Brown  | 12    | 2     | 14    |
| Total  | 43    | 33    | 76    |

5. Calculate the **expected values** in each cell using the formula

$$\text{expected value} = \frac{(\text{row total}) \times (\text{column total})}{(\text{table total})}$$

**(Continuation of example 57.1.)**
The expected values are the numbers in parentheses.

|        | Small       | Large      | Total |
|--------|-------------|------------|-------|
| Red    | 10 (9.56)   | 6 (7.33)   | 16    |
| Orange | 3 (7.76)    | 10 (5.96)  | 13    |
| Yellow | 6 (6.57)    | 5 (5.04)   | 11    |
| Green  | 6 (5.97)    | 4 (4.58)   | 10    |
| Blue   | 6 (7.17)    | 6 (5.5)    | 12    |
| Brown  | 12 (8.36)   | 2 (6.42)   | 14    |
| Total  | 43          | 33         | 76    |

6. Calculate the **chi-squared statistic**

$$\chi^2 = \sum \frac{(O - e)^2}{e}$$

summing over all cells in the table. Here $O$ is observed value, and $E$ is expected value.

FAQ 57. What is a Chi Squared Homogeneity Test?

**(Continuation of example 57.1.)**

$$\chi^2 = \frac{(10-9.56)^2}{9.56} + \frac{(3-7.76)^2}{7.76} + \frac{(6-6.57)^2}{6.57} + \frac{(6-5.97)^2}{5.97}$$
$$+ \frac{(6-7.17)^2}{7.17} + \frac{(12-8.36)^2}{8.36} + \frac{(6-7.33)^2}{7.33} + \frac{(10-5.96)^2}{5.96}$$
$$+ \frac{(5-5.04)^2}{5.04} + \frac{(4-4.58)^2}{4.58} + \frac{(6-5.5)^2}{5.5} + \frac{(2-6.42)^2}{6.42}$$
$$= 11.283$$

7. Calculate the **degrees of freedom**:

$$df = \text{categories} - 1$$

   **(Continuation of example 57.1.)** There are six categories in the table, so the number of degrees of freedom is $df = 6 - 1 = 5$.

8. **Determine the p-value** (e.g., using table reftable-chi-squared).

   **(Continuation of example 57.1.)**
   a) Above we found that $\chi^2 = 11.283$ and $df = 5$.
   b) We examine row 5 of table A.3, and find two values that bracket 11.28 are 11.07 and 12.83. Hence $11.07 < \chi^2 < 12.83$.
   c) Following the columns containing 11.07 and 12.83 to the top row of the table we find bracketing p-values of 0.025 and 0.05, so that

   $$0.025 < \text{p-value} < 0.05$$

9. **Interpret the result by comparing the p-value with $\alpha$.**

   a) If $p \leqslant \alpha$, we reject $H_0$ and accept $H_A$. We would conclude that the populations have different distributions.
   b) If $p > \alpha$ then there is not enough evidence to reject $H_A$. We do not make any conclusion.

   **(Continuation of example 57.1.)** Since the p-value is smaller than the stated value of $\alpha = 0.05$, we reject $H_0$ and accept $H_A$. In terms of the language of this particular example, it means that the color distribution of the candy in the two different sized bags is different.

**Technology Example 57.2 (TI-84).** Repeat the calculation with the data from example 57.1 using a TI-84.

1. Enter the observed data into matrix **[A]**. (You can use any matrix, the default is to use matrix **[A]**.)

---

a) To enter matrix data, click on [2nd] [matrix]
b) Scroll to [EDIT] ⟩ [A] and click [enter]
c) On the first line enter the matrix size **6** (number of rows) by **2** (number of columns). A template will appear.
d) Enter the observed values in the appropriate locations. Press enter after entering each value.

2. After all the observed values have been entered into matrix **[A]** click [stat] ⟩ [TESTS].

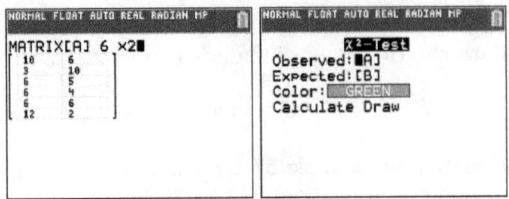

3. Scroll to [$x^2$-Test] and press [enter]. This test is on line **C** (11) and you may have to scroll off the bottom of the screen to find it.

   a) For **Observed** select matrix **[A]**. (The specific use of matrix **[A]** is not required; you can use any matrix, and enter it here.)
   b) For **Expected** select matrix **[B]**. The matrix you select here will be overwritten with the values calculated by the $x^2$ test. Note that you should not calculate the expected values in advance. (The specific use of matrix **[B]** is not required; you can use any matrix, besides the matrix you use for the observed values, and enter it here.)
   c) Scroll to **Calculate** and press [enter].

4. The results of the $chi^2$ calculate appear on the $\chi^2$-**Test** screen. We see that $\chi^2 = 11.2825424$ and $p = 0.0460570183$. This agrees with our previous conclusion that $p < 0.05$, but is more precise.

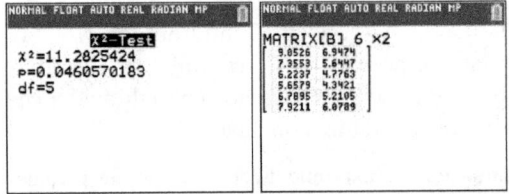

5. The expected values are also computed and placed in matrix **[B]**. To see them, select [2nd] ⟩ [matrix] ⟩ [EDIT] ⟩ [B]. If you chose a different matrix than **[B]** for the expected values during the input step, then you should look in that matrix instead of **[B]** for the expected values.

# Exercises

1. Suppose the following are two different samples of test scores from students. Sample 1: {61, 70, 83, 71, 86, 78, 65, 84, 100, 73}; Sample 2: {79, 69, 85, 88, 70, 74, 88, 66, 67, 61} Use a $\chi^2$ homogeneity test to

test the likelihood that the samples came from the same class. (Ans: $\chi^2 = 18.748$, $df = 9$, $p = 0.02743$)

2. Suppose the following are two different samples of test scores from students. Sample 1: {61, 70, 83, 71, 86, 78, 65, 84, 100, 73}; Sample 2: {82, 71, 79, 79, 76, 79, 82, 83, 76, 71} (Ans: $\chi^2 = 9.4811$; $df = 9$; $p = 0.3941$)

# FAQ 58. What is a Chi Squared Goodness of Fit Test?

The **chi squared goodness of fit test** is used to determine whether or not a data set can be predicted from a particular distribution. For example, we me ask if the data is evenly distributed.

**Example 58.1.** Are the colors of candy in a bag of Skittles all distributed evenly?

The chi squared goodness of fit test sums up the $(O - E)^2/E$ values in the same manner as the chi square test for independence (chapter 56). The only difference is that instead of having a table with $r$ rows and $c$ columns, we only have $n$ data values for comparison. Consequently the number of degrees of freedom is calculated differently. In this case, it is the df=$n-1$. This is analogous to the $\chi^2$ test for homogeneity, where the number of degrees of freedom is equal to the number of categories minus one.

This is how we perform a goodness of fit test.

1. **List (and sum) the distribution.**

   (**Continuation of example 58.1.**) To get a sample of Skittles we go to the store a buy a bag. We pretend that this represents a random sample of all of the Skittles in the known universe.

   | | Red | Blue | Green | Orange | Purple | Magenta | Sum |
   |---|---|---|---|---|---|---|---|
   | Observed Count | 10 | 3 | 18 | 2 | 11 | 17 | 61 |

2. **Make a hypothesis** about the expected distribution.

   (**Continuation of example 58.1.**) We are testing against the hypothesis that the skittles are evenly distributed, which means they all have the same proportion:

   $H_0$: the candy colors are evenly distributed

$H_A$: the candy colors are not evely distributed

---

## Chi Squared Test for Goodness of Fit

To test if a data set $x_1, x_2, \ldots, x_n$ matches predicted (or expected) values $e_1, e_2, \ldots, e_n$ form the sum

$$\chi^2 = \sum \frac{(x_i - e_i)^2}{e_i}$$

The sum $\chi^2$ follows a chi-squared distribution with $n - 1$ degrees of freedom.

---

3. Based on your null hypothesis, **predict the expected values.**

   (**Continuation of example 58.1.**)   If the candies are evenly distributed, we would expect the same number of candies in each color. Since there are six colors and there are 61 colors, each color would have an expected value of 61/6=10.17 (rounded to two decimal places). More precisely, the proportion $p$ in each color would be 1/6,

   $$H_0 : p = 1/6$$
   $$H_A : p \neq 1/6$$

   and therefore, under the assumption that $H_0$ is true, the expected value in each category is 10.17.

   | | Red | Blue | Green | Orange | Purple | Magenta | Sum |
   |---|---|---|---|---|---|---|---|
   | Expected Count | 10.17 | 10.17 | 10.17 | 10.17 | 10.17 | 10.17 | 61 |

4. **Form the O(E) table.**

   (**Continuation of example 58.1.**)

   | | Red | Blue | Green | Orange | Purple | Magenta |
   |---|---|---|---|---|---|---|
   | O(E) | 10 (10.17) | 3(10.17) | 18 (10.17) | 2(10.17) | 11 (10.17) | 17 (10.17) |

5. **Calculate the sum** of $(O - E)^2/E$ for the table

   $$\chi^2 = \sum_{\text{all data}} \frac{(\text{observed value - expected value})^2}{(\text{expected value})}$$

---

FAQ 58.  What is a Chi Squared Goodness of Fit Test?

**(Continuation of example 58.1.)**

$$\chi^2 = \frac{(10-10.17)^2}{10.17} + \frac{(3-10.17)^2}{10.17} + \frac{(18-10.17)^2}{10.17} + \frac{(12-10.17)^2}{10.17} +$$
$$\frac{(11-10.17)^2}{10.17} + \frac{(17-10.17)^2}{10.17}$$
$$=0.00284 + 5.055 + 6.028 + 6.563 + 0.0677 + 4.587$$
$$=22.31$$

6. Determine the **degrees of freedom**

$$df = (\text{number of columns}) - 1.$$

(**Continuation of example 58.1.**) Since there are 6 rows, there df $= 6-1=5$.

7. **Determine the p value,** either from software or table A.3.

(**Continuation of example 58.1.**) $p = 0.000457$ from software, or $p < .001$ from table A.3. Therefore

a) The low $p$ value indicates that if the sampling were repeated many times, and if the null hypothesis $H_0$ were true, that a distribution this far from evenly distributed would only occur 1 out of every 2188 times $(1/.000457 = 2188)$ by random chance less than 1 out of 1000 if use the table read).
b) The observed variation is extremely rare and unlikely to occur.
c) Our observation is evidence *against the null hypotheses* because it is rare, and because the null hypothesis is true.
d) Our observation is *statistically significant* because the $p$ value is very low.
e) We reject $H_0$, accept $H_A$, and conclude that the colors are unevenly distributed.

**Example 58.2.** Another candy manufacturer distributes Blobbo Candy Pieces. They advertise that they use the following mixture:

|         | Red | Blue | Green | Orange | Purple | Magenta |
|---------|-----|------|-------|--------|--------|---------|
| Percent | 15  | 5    | 30    | 5      | 20     | 25      |

You think that maybe the candy you used in the previous example was really Blobbos and not Skittles. So you decide to test the following hypothesis:

$$H_0: \text{The distribution of candy is as advertised by Blobbos}$$

$H_A$: The distribution is not as advertised

The previous sample size was 61. To get the expected amounts we multiply predicted percents by $61/100 = 0.61$:

|  | Red | Blue | Green | Orange | Purple | Magenta |
|---|---|---|---|---|---|---|
| Expected | .61× 15 | .61×5 | .61×30 | .61×5 | .61×20 | .61×25 |

Here are the resulting expected values:

|  | Red | Blue | Green | Orange | Purple | Magenta |
|---|---|---|---|---|---|---|
| Expected | 9.15 | 3.05 | 18.3 | 3.05 | 12.2 | 15.25 |

Next we form the O(E) table:

|  | Red | Blue | Green | Orange | Purple | Magenta |
|---|---|---|---|---|---|---|
| O(E) | 10 (9.15) | 3 (3.05) | 18 (18.3) | 2 (3.05) | 11(12.2) | 17 (15.25) |

Here we compute $\chi^2 = \sum \dfrac{(O-E)^2}{E}$:

$$\chi^2 = \frac{(10-9.15)^2}{9.15} + \frac{(3-3.05)^2}{3.05} + \frac{(18-18.3)^2}{18.3} + \frac{(2-3.05)^2}{3.05}$$
$$+ \frac{(11-12.2)^2}{12.2} + \frac{(17-15.25)^2}{15.25}$$
$$=0.765$$

Table A.3 gives $p > 0.25$ (table) using $df = 5$. Thus the data does not provide good evidence against $H_0$, as the event may occur at least one time in four by random chance chance when $H_0$ is true.[1] Thus there is not sufficient evidence to reject the claim that the candy has the same distribution as advertised by Blobbos.

**Technology Example 58.3 (TI-84).** Repeat example 58.2 using a TI-84.

This can be done using the $\chi^2$**GOF-Test** function of the calculator.

1. Enter the observed values $\{10, 3, 18, 2, 11, 17\}$ into list **L1**. (To enter data into a list hit $\boxed{\text{stat}}$ followed by $\boxed{\text{EDIT}} \gg \boxed{\text{edit}}$ $\boxed{\text{enter}}$. Navigate to the desired list with the arrow keys. To clear any existing data, navigate to the list label and then $\boxed{\text{clear}}$ $\boxed{\text{enter}}$. Then navigate to the first entry of the list. Enter each data item followed by the $\boxed{\text{enter}}$.)
2. Enter the expected values $\{9.15, 3.05, 18.3, 3.05, 12.2, 15.25\}$ into list **L2**.

---

[1] In fact it may occur much more frequently than that because the p-value is off the left-end of the table and is therefore quite likely much greater than 0.25.

FAQ 58. What is a Chi Squared Goodness of Fit Test?

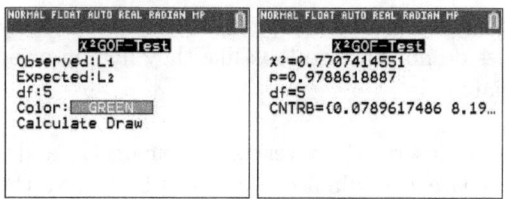

3. Open the $\chi^2$ goodness of fit menu. Hit [stat], then [TESTS] $\rangle$ $\chi^2$GOF-Test [enter]. (This is on line **D** (line 15) so you will have scroll down past the bottom of your scree to locate it.)

4. On the $\chi^2$**GOF-Test** input menu,

   a) For **Observed** enter **L1** (If you chose a different list for your observed data set, enter it instead of **L1**).

   b) For **Expected** enter **L2** (If you chose a different list for your expected values, enter it instead of **L2**).

   c) For **df** enter **5** since there are 5 degrees of freedom for this example.

   d) Navigate to **Calculate** and hit [enter].

5. The result will be displayed on the $\chi^2$**GOF-Test** output screen. We see that $\chi^2 = 0.7707414551$ and $p = 0.9788618887$. The $\chi^2$ value is consistent with our previous calculation. The large $p$ value confirms our suspicion (expressed in the footnote) that our earlier lower bound on the p-value was too low.

---

# FAQ 59. What is a Scatter Plot?

Suppose we have a collection of data pairs $(x_1, y_1)$, $(x_2, y_2)$, $(x_3, y_3)$, ..., $(x_n, y_n)$. We can visualize this type of data using a scatter plots. The points may be represented by small dots or symbols such as squares, triangles, circles, etc. (fig. 59.1).

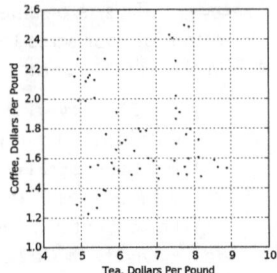

Fig. 59.1. World coffee vs. tea price, per pound (2011 to 2016).

- ▸ We call the $x$ variable the **explanatory variable**
- ▸ We call the $y$ variable the **response variable**.

If there is a clear pattern in the graph, we say that they are **associated**. The variables do not have to look like they line on or near a straight line to be associated.

Association is described in terms of **strength** and **direction**. The **strength of the association** describes how close the actual observed data matches the qualitative descriptor of the pattern.

- ▸ **Strongly associated**: there is a clear pattern or geometric shape, such as a line or curve, with most of the data matching the pattern (e.g., fig 59.2, center; there is a strong, curved pattern of association).
- ▸ **Weekly associated**: there is a pattern but a large fraction of the data does not match the pattern (e.g., fig 59.2, left; there is a weak, linear association).
- ▸ **Clustered**: the data appears to be grouped into specific regions of the plot (e.g., fig 59.2, right).
- ▸ **Not associated:** there is no clear pattern (e.g., fig. 59.1).

If the patter is generally increasing or generally decreasing as $x$ increases, then we assign a direction to the association. More specifically, two variables are said to be

- ▸ **positively associated** (fig. 59.3) if one variable increases when the other variable increases

- ▸ **negatively associated** if one variable decreases when the other one increases.

Figure 59.2.: Different patterns of association. Left to right: linear; curved; clusterd.

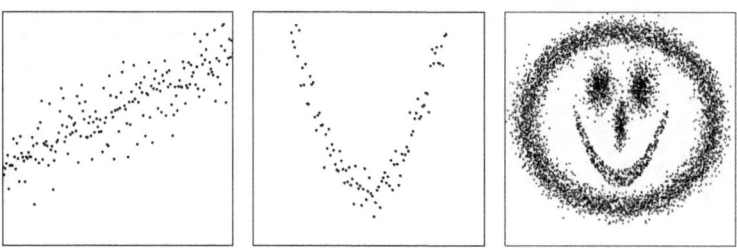

Figure 59.3.: Comparison of strong and weak association, and positive and negative association.

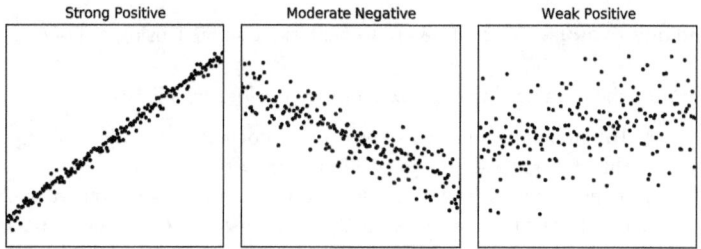

**Technology Example 59.1 (R).** Suppose that the Math SAT scores of eight students are {353, 370, 4124, 470, 490, 510, 520, 610}. Suppose further that the grade point average (GPA) of the corresponding students, in order, during their first semester in college is {2.7, 2.8, 2.5, 3.2, 3.3 3.5, 3.4, 3.7}. Using R, make a scatter plot of GPA as a function of SAT score.

We can do this using the **plot (x,y)** function. Here **x** is a vector is $x$ data values and and**y** is a vector containing the $y$ data values.

```
> sats=c(353, 370, 412, 470, 490, 510, 520, 610)
> gpas=c(2.7, 2.8, 2.5, 3.2, 3.3, 3.5, 3.4, 3.7)
> plot(sats,gpas)
```

A plot will pop up in a new window automatically. In the RStudio console it will appear in the **Plots** window.

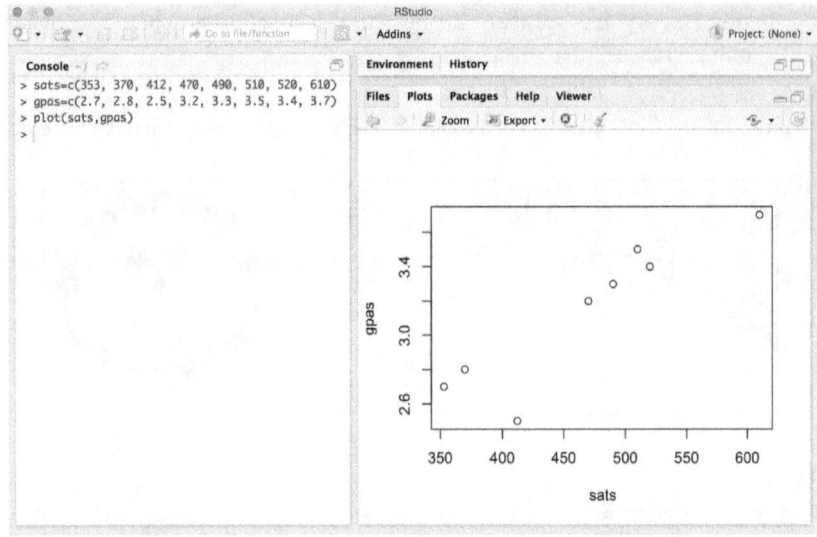

**Technology Example 59.2 (TI-84).** Repeat exercise 59.1 using a Ti-84.

1. Enter the $x$ values into list **L1** and the $y$ values into list **L2**.

   a) To enter data into a list, you must open the statistics editing mode. Hit [stat] followed by [edit] ⟩ [EDIT] and [enter]

   b) If the columns for list **L1** and **L2** contain data and you want to clear this data first, navigate to the column header (the label **L1** or **L2**) with the arrow keys. The press [clear] [enter] to clear the entire list.

   c) If the columns for list **L1** and **L2** are empty you may enter the numbers one at a time, pressing [enter] after each value.

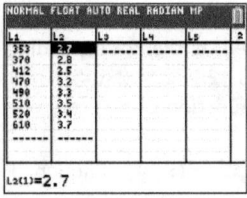

2. Select plotting mode [2nd] [y=] [1] and then set up the plotting parameters.

   a) Select [Plot1] ⟩ [On].
   b) Select scatter plot from **Type** (this is the first plot icon).
   c) Select **L1** for **Xlist**.
   d) Select **L2** for **Ylist**.
   e) Select your choice of plotting symbol for **Mark**.
   f) Select your choice of plotting color for **Color**

---

FAQ 59.   What is a Scatter Plot?

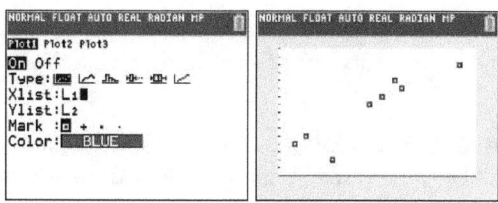

3. To display the plot press [graph]. If the plot appears blank or is very small (only appears in a small part of the screen) press [zoom] [9] to zoom to an appropriate scale.

# Exercises

Make scatter plots of each of the following data sets. Identify the type, strength, and if appropriate, the direction of the association for each.

1. (23, 19.9), (54, 12.8), (84, 35.0), (28, 9.8), (88, 38.1), (22, 6.5), (11, 1.0), (90, 39.7), (11, 11.1), (64, 26.7), (40, 22.3), (85, 29.2), (12, 6.4), (53, 22.1), (51, 12.7)

2. (90, -68.4), (10, -9.5), (31, -21.4), (43, -27.2), (72, -64.2), (33, -32.1), (72, -46.7), (75, -49.4), (82, -70.5), (57, -49.0), (26, -21.8), (34, -34.4), (61, -52.5), (13, -14.0)

3. (22, 11.9), (21, 15.0), (39, 8.6), (21, 7.2), (19, 11.2), (58, 29.5), (80, 35.8), (29, 18.5), (40, 25.0), (14, -0.6), (89, 33.9), (78, 22.2), (19, 12.4), (71, 27.0), (26, 17.1)

4. (87, 28.0), (28, 7.3), (26, 14.8), (34, 14.4), (21, 15.5), (89, 36.4), (73, 24.7), (66, 18.7), (47, 20.3), (81, 20.0), (20, 9.1), (33, 15.2), (70, 18.1), (66, 30.8), (89, 32.6)

5. (57, -13.5), (69, -3.1), (50, 21.4), (90, 9.9), (62, 30.2), (137, -18.1), (29, 27.3), (55, 21.5), (32, 24.1), (59, 2.0), (71, 4.8), (29, 13.2), (92, 36.2), (44, 22.6), (140, 0.6), (132, 16.1), (84, -18.2)

# FAQ 60. What is the Correlation?

**The correlation** $r$ is a number between -1 and 1. It gives a quantitative measure of the goodness of the description of the association, but only when that description is linear.

▸ Correlation is only meaningful if the association is linear.
▸ Correlation only describes relations between quantitative variables. It is undefined for categorical variables.

**Example 60.1.** You read an article that says there is a correlation between gender and salary in the tech industry. How do you know that this wrong?

You know that this statement is mathematically incorrect because gender is categorical variable and correlation is not defined for categorical variables. What the author probably means is that the mean salary of men is different than the mean salary of women, but the use of the word correlation in this case is incorrect.

**Correlation and Association.** Correlation is related to association but they are not the same thing. Usually when variables are highly correlated they show a strong association when plotted. However, the opposite is not necessarily true. Two variables can be very strongly associated but not highly correlated. This is because the **correlation only measures the part of the relationship that is linear,** and if the relationship between the two variables is nonlinear, the correlation could be small (or even zero).

## The Sign of the Correlation is Related to Slope.

- When $r > 0$, $y$ tends to increase as $x$ increases.
- When $r < 0$, $y$ tends to decrease as $x$ decreases.

Thus

- positive correlation indicates a positive association, and
- negative correlation indicates a negative association;

but **only when the pattern is linear.**

**The Magnitude of the Correlation.** The correlation always has magnitude less than one:

$$-1 \leqslant r \leqslant 1$$

If you calculate a correlation that is greater than 1 or less than -1, you have made a mistake.

**Verify that the Pattern is Linear.** The correlation has little or no meaning when the relationship between the $x$ and $y$ data is not linear.

**Correlations Close to 1 or -1.** If the correlation is close to 1 or -1 (and the pattern is linear) most of the variation in the data can be explained by a line. Further, for linear patterns:

- For $r$ is close to 1 there is a strong positive linear association.
- For $r$ is close to -1 there is strong negative linear association.

FAQ 60. What is the Correlation?

**Coefficient of Determination.** When the pattern is linear, the value of $r^2$, gives the proportion of variation in $y$ with $x$ that is described by the fit.

**The Correlation is Symmetric.** If you exchange *all* of the $x$ values with *all* of the $y$ values, you will get the same value for the correlation between $x$ and $y$.

**The Correlation is Dimensionless.** The value of $r$ will be the same no matter what units you use for your data. However, every $x$-value must have the same units as every other $x$-value, and every $y$-value must have the same units as every other $y$-value.

**Correlation Does Not Imply Causation.** Causation can only be demonstrated by a controlled experiment. This is discipline-specific and beyond the scope of this book.

**Example 60.2.** The following table compares the number of Starbucks in the world from 1999 through 2009 with the enrollment (in thousands of students) in the California State University system in January of the same year.

| Year | Starbucks Stores, Worldwide | CSU Enrollment Thousands |
|------|------|------|
| 1999 | 2498 | 270 |
| 2000 | 3501 | 278 |
| 2001 | 4709 | 294 |
| 2002 | 5886 | 304 |
| 2003 | 7225 | 301 |
| 2004 | 8569 | 303 |
| 2005 | 10241 | 308 |
| 2006 | 12440. | 325 |
| 2007 | 15011 | 338 |
| 2008 | 16680 | 343 |
| 2009 | 16635 | 320 |

The relationship between the number of Starbucks (the explanatory variable) and CSU enrollment (the response variable) is shown in fig. 60.1. There appears to be an association between the two variables, and a high correlation ($r = 0.93$). However, it seems very unlikely that either variable directly affects the other.

Figure 60.1.: Correlation does not imply causation. Plot of CSU enrollment versus number of Starbucks. The line is the least squares linear regression (ch. 62). The correlation coefficient is $r = 0.93$ (ch. **??**

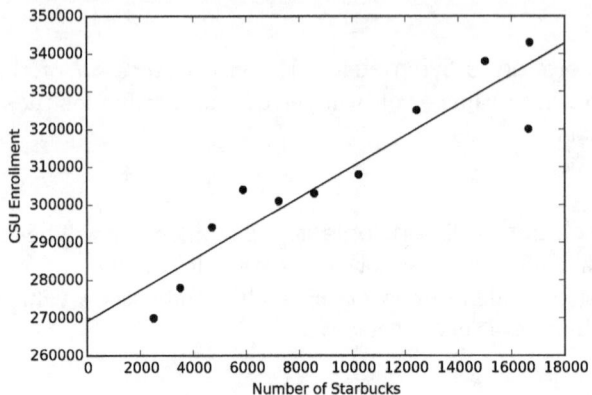

# FAQ 61. How Do I Find the Correlation?

In chapter 60 we discussed the general properties of the correlation. We now turn to the procedure for actually calculating the correlation. The **correlation** $r$ between $x$ and $y$ in a sample of $n$ **data pairs**

$$(x_1, y_1), (x_2, y_2), \ldots, (x_n, y_n)$$

is given by the equation

$$r = \frac{1}{n-1} \sum_{i=1}^{n} \left( \frac{x_i - \bar{x}}{s_x} \right) \left( \frac{y_i - \bar{y}}{s_y} \right)$$

where $\bar{x}$ and $s_x$ are the sample mean and sample standard deviation of $x_1, x_2, \ldots, x_n$ (ch. 13, 15); and $\bar{y}$ and $s_y$ are the sample standard deviation of $y_1, y_2, \ldots, y_n$. Because this formula is so messy, the correlation is rarely calculated by hand. Statistical software, a spreadsheet, or (as a last resort) a calculator is normally used. When it is necessary to do the calculation manually, extreme patience is required because of the large number of calculations involved.

**Example 61.1.** Find the correlation between the x values and the y values in the following set of data pairs:

$$(1.1, 2.2), (1.9, 2.3), (2.1, 2.42), (2.8, 2.49), (3.7, 2.83), (4.5, 2.75)$$

More often than not, the data is written in a table rather than a list of pairs. The table may be written either horizontally or vertically.

**(Continuation of example 61.1.)** We can write the same data pairs in a horizontal table in this way:

| x | 1.1 | 1.9 | 2. | 2.8 | 3.7 | 4.5 |
|---|-----|-----|-----|------|------|------|
| y | 2.2 | 2.3 | 2.42 | 2.49 | 2.83 | 2.75 |

To find the correlation:

1. Find $\overline{x}$, the **mean of the explanatory variable data**, i.e., of $x_1, \ldots, x_n$ (ch. 13),

$$\overline{x} = \frac{x_1 + x_2 + \cdots + x_n}{n}$$

**(Continuation of example 61.1.)**

$$\overline{x} = \frac{1.1 + 1.9 + 2. + 2.8 + 3.7 + 4.5}{6} = 2.68.$$

2. Find $s_x$, the **standard deviation of the explanatory variable data**, i.e., of $x_1, \ldots, x_n$ (ch. 15),

$$s_x^2 = \frac{1}{n-1} \sum_{i=1}^{n} (x_i - \overline{x})$$

where $\overline{x}$ is the **mean of the explanatory variable data** (step 1).

**(Continuation of example 61.1.)**

$$s_x^2 = \frac{1}{n-1} \sum_{i=1}^{n} (x_i - \overline{x})$$

$$= \frac{1}{6-1} \big[ (1.1 - 2.68)^2 + (1.9 - 2.68)^2 + (2 - 2.68)^2 + (2.8 - 2.68)^2$$

$$+ (3.7 - 2.68)^2 + (4.5 - 2.68)^2 \big]$$

$$= 1.59$$

Hence the standard deviation is $s_x = \sqrt{s_x^2} = \sqrt{1.59} = 1.26$

3. Find $\bar{y}$, the **mean of the the response variable data**, i.e., of $y_1, \ldots, y_n$ (ch. 13),

$$\bar{y} = \frac{y_1 + y_2 + \cdots + y_n}{n}$$

**(Continuation of example 61.1.)**

$$\bar{y} = \frac{2.2 + 2.3 + 2.42 + 2.49 + 2.83 + 2.75}{6} = 2.50.$$

4. Find $s_y$, the **standard deviation of the response variable data**, i.e., of $y_1, \ldots, y_n$ (ch. 14),

$$s_y^2 = \frac{1}{n-1} \sum_{i=1}^{n} (y_i - \bar{y})$$

Here $\bar{y}$ is the **mean of the response variable data** (step 3).

**(Continuation of example 61.1.)**

$$s_y^2 = \frac{1}{n-1} \sum_{i=1}^{n} (y_i - \bar{y})$$

$$= \frac{1}{6-1} [(2.2 - 2.5)^2 + (2.3 - 2.5)^2 + (2.42 - 2.5)^2 + (2.49 - 2.5)^2$$

$$+ (2.83 - 2.50)^2 + (2.75 - 2.5)^2] =$$

$$0.062$$

Thus the standard deviation is $s_y = \sqrt{s_y^2} = \sqrt{0.062} = 0.025.$

5. Write out the $n$ terms of the formula for the correlation

$$r = \frac{1}{n-1} \sum_{i=1}^{n} \left[ \frac{x_i - \bar{x}}{s_x} \right] \left[ \frac{y_i - \bar{y}}{s_y} \right]$$

**(Continuation of example 61.1.)**   Since $n = 6$ (and hence $n - 1 = 5$), the formula for the correlation is

$$r = \frac{1}{5} \left\{ \left[ \frac{x_1 - \bar{x}}{s_x} \right] \left[ \frac{y_1 - \bar{y}}{s_y} \right] + \left[ \frac{x_2 - \bar{x}}{s_x} \right] \left[ \frac{y_2 - \bar{y}}{s_y} \right] + \left[ \frac{x_3 - \bar{x}}{s_x} \right] \left[ \frac{y_3 - \bar{y}}{s_y} \right] + \right.$$

$$\left. \left[ \frac{x_4 - \bar{x}}{s_x} \right] \left[ \frac{y_4 - \bar{y}}{s_y} \right] + \left[ \frac{x_5 - \bar{x}}{s_x} \right] \left[ \frac{y_5 - \bar{y}}{s_y} \right] + \left[ \frac{x_6 - \bar{x}}{s_x} \right] \left[ \frac{y_6 - \bar{y}}{s_y} \right] \right\}$$

6. Fill in the values of $s_x$ and $s_y$ with the values determined above in steps 2 and 4. Note that these same two values are used in each denominator.

**(Continuation of example 61.1.)**   Since $s_x = 1.26$ and $s_y = 0.25$, we replace *every occurrence* of $s_x$ with the number 1.26, and *every occurrence* of $s_y$ with the number 0.25.

$$r = \frac{1}{5}\left\{\left[\frac{x_1 - \overline{x}}{1.26}\right]\left[\frac{y_1 - \overline{y}}{0.25}\right] + \left[\frac{x_2 - \overline{x}}{1.26}\right]\left[\frac{y_2 - \overline{y}}{0.25}\right] + \left[\frac{x_3 - \overline{x}}{1.26}\right]\left[\frac{y_3 - \overline{y}}{0.25}\right] + \right.$$
$$\left.\left[\frac{x_4 - \overline{x}}{1.26}\right]\left[\frac{y_4 - \overline{y}}{0.25}\right] + \left[\frac{x_5 - \overline{x}}{1.26}\right]\left[\frac{y_5 - \overline{y}}{0.25}\right] + \left[\frac{x_6 - \overline{x}}{1.26}\right]\left[\frac{y_6 - \overline{y}}{0.25}\right]\right\}$$

7. Fill in the values of $\overline{x}$ and $\overline{y}$ with the values determined above in steps 1 and 3.

**(Continuation of example 61.1.)**   Since $\overline{x} = 2.68$ and $\overline{y} = 2.50$, we replace *every occurrence* of $\overline{x}$ with the number 2.68, and *every occurrence* of $\overline{y}$ with the number 2.50.

$$r = \frac{1}{5}\left\{\left[\frac{x_1 - 2.68}{1.26}\right]\left[\frac{y_1 - 2.5}{0.25}\right] + \left[\frac{x_2 - 2.68}{1.26}\right]\left[\frac{y_2 - 2.5}{0.25}\right] + \right.$$
$$\left[\frac{x_3 - 2.68}{1.26}\right]\left[\frac{y_3 - 2.5}{0.25}\right] + \left[\frac{x_4 - 2.68}{1.26}\right]\left[\frac{y_4 - 2.5}{0.25}\right] +$$
$$\left.\left[\frac{x_5 - 2.68}{1.26}\right]\left[\frac{y_5 - 2.5}{0.25}\right] + \left[\frac{x_6 - 2.68}{1.26}\right]\left[\frac{y_6 - 2.5}{0.25}\right]\right\}$$

8. Fill in all of the $x_i$ and $y_i$ values from the original data table.

**(Continuation of example 61.1.)**
Here are the original data values and the index ($i$) values from 1 to 6:

| $i$ | 1 | 2 | 3 | 4 | 5 | 6 |
|---|---|---|---|---|---|---|
| $x_i$ | 1.1 | 1.9 | 2. | 2.8 | 3.7 | 4.5 |
| $y_i$ | 2.2 | 2.3 | 2.42 | 2.49 | 2.83 | 2.75 |

Plugging all the numbers into the formula for $r$ gives

$$r = \frac{1}{5}\left\{\left[\frac{1.1 - 2.68}{1.26}\right]\left[\frac{2.2_1 - 2.5}{0.25}\right] + \left[\frac{1.9 - 2.68}{1.26}\right]\left[\frac{2.3 - 2.5}{0.25}\right] + \right.$$
$$\left[\frac{2 - 2.68}{1.26}\right]\left[\frac{2.42 - 2.5}{0.25}\right] + \left[\frac{2.8 - 2.68}{1.26}\right]\left[\frac{2.49 - 2.5}{0.25}\right] +$$
$$\left.\left[\frac{3.7 - 2.68}{1.26}\right]\left[\frac{2.83 - 2.5}{0.25}\right] + \left[\frac{4.5 - 2.68}{1.26}\right]\left[\frac{2.75 - 2.5}{0.25}\right]\right\}$$
$$= \frac{1}{5}[(-1.27)(-1.50) + (-.63)(-.80) + (-47)(-32) + (.09)(-.03)+$$

$$= \frac{1.52 + 0.5 + 0.15 - 0.0 + 1.09 + 1.46}{5}$$

$(.81)(1.34) + (1.45)(1.01)]$

$= 0.944$

**Technology Example 61.2 (TI-84).** Find the correlation of the $x$ and $y$ given in example 61.1 using a TI-84.

1. Place the calculator in diagnostic mode. In general, you will only have to do this when the calculator is new, or after you have reset it. Diagnostic mode allows it to display additional results. Locate the [catalog] button on your calculator, which is above the number [0]. Press [2nd] [catalog] to open the catalog.

2. Inside the catalog, scroll down to **DiagnosticOn** and click [enter] twice. This places the calculator in diagnostic mode.

 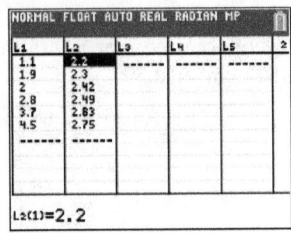

3. Enter all of the $x$ values into **L1** and all of the $y$ values into **L2**. The values must be entered in the same order as they are in the data.

   a) To enter edit mode, press [stat], then [EDIT] [1].
   b) Scroll to each list and type in the numbers one at a time, following each number with the [enter] key. Put corresponding $(x, y)$ pairs on the same horizontal line.
   c) When you are done entering data, press [2nd] [quit].

4. Press [stat], scroll to [CALC] ≫ [LinReg(ax+b)], and press [enter].

5. For the **Xlist**, select **L1**; for the **Ylist**, select **L2**; then select [calculate].

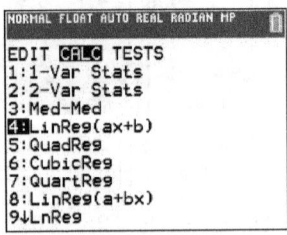 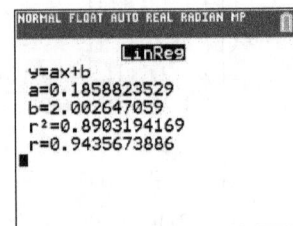

The correlation is the number .9435673886 labeled as $r$ on the calculator display.

**Technology Example 61.3 (R).** Find the correlation between the number of Starbucks franchises and the enrollment at CSU (see example 60.2) using R. Demonstrate (a) that the result is independent of units; and (b), that it does not depend on the order of the arguments.

We can do this with the R function `cor(x, y)`, where `x` and `y` are arrays that contain the values of the $x$ and $y$ data.

```
> starbucks=c(2498,3501,4709,5886,7225,8569,10241,
  12440,15011,16680,16635)
> enrollment=c(270,278,294,304,301,303,308,325,338,
  343,320)
> cor(starbucks,enrollment)
[1] 0.9312839
```

If we express the enrollment as individuals rather than thousands of students, the answer is unchanged:

```
> starbucks=c(2498,3501,4709,5886,7225,8569,10241,
  12440,15011,16680,16635)
> e=c(270000,278000,294000,304000,301000,303000,
  308000,325000,338000,343000,320000)
> cor(starbucks,e)
[1] 0.9312839
```

Furthermore, if we reverse the order of the arguments, the correlation is unchanged:

```
> cor(e,starbucks)
[1] 0.9312839
```

**Technology Example 61.4 (Spreadsheet).** Find the correlation of the Starbucks data using a spreadsheet.

1. Type the $x$ and $y$ data into two columns of the spreadsheet.
2. Type "=**correl(**" and select the $x$ and $y$ data with the mouse. There must be a comma between the $x$ data selection and the $y$ data selection. The syntax is **correl(x, y)**.

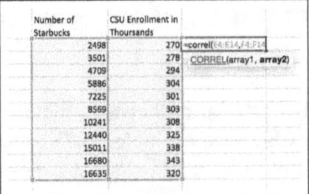

3. Hit enter and the answer appears in the same cell where you typed `=correl`.

# Exercises

Find the correlation between $x$ and $y$ for each of the following data sets that were plotted in chapter 59.

1. (23, 19.9), (54, 12.8), (84, 35.0), (28, 9.8), (88, 38.1), (22, 6.5), (11, 1.0), (90, 39.7), (11, 11.1), (64, 26.7), (40, 22.3), (85, 29.2), (12, 6.4), (53, 22.1), (51, 12.7) (ans: $\bar{x} = 47.733$, $\bar{y} = 19.553$, $s_x = 29.463$, $s_y = 11.814$, $r = 0.901$)

2. (90, -68.4), (10, -9.5), (31, -21.4), (43, -27.2), (72, -64.2), (33, -32.1), (72, -46.7), (75, -49.4), (82, -70.5), (57, -49.0), (26, -21.8), (34, -34.4), (61, -52.5), (13, -14.0) (ans: $\bar{x} = 49.929$, $\bar{y} = -40.079$, $s_x = 26.146$, $s_y = 19.335$, $r = -0.955$)

3. (22, 11.9), (21, 15.0), (39, 8.6), (21, 7.2), (19, 11.2), (58, 29.5), (80, 35.8), (29, 18.5), (40, 25.0), (14, -0.6), (89, 33.9), (78, 22.2), (19, 12.4), (71, 27.0), (26, 17.1) (ans: $\bar{x} = 41.733$, $\bar{y} = 18.313$, $s_x = 26.196$, $s_y = 10.079$, $r = 0.852$)

4. (87, 28.0), (28, 7.3), (26, 14.8), (34, 14.4), (21, 15.5), (89, 36.4), (73, 24.7), (66, 18.7), (47, 20.3), (81, 20.0), (20, 9.1), (33, 15.2), (70, 18.1), (66, 30.8), (89, 32.6) (ans: $\bar{x} = 55.333$, $\bar{y} = 20.393$, $s_x = 26.378$, $s_y = 8.227$, $r = 0.843$)

5. (57, -13.5), (69, -3.1), (50, 21.4), (90, 9.9), (62, 30.2), (137, -18.1), (29, 27.3), (55, 21.5), (32, 24.1), (59, 2.0), (71, 4.8), (29, 13.2), (92, 36.2), (44, 22.6), (140, 0.6), (132, 16.1), (84, -18.2) (ans: $\bar{x} = 72.471$, $\bar{y} = 10.412$, $s_x = 35.945$, $s_y = 16.353$, $r = -0.399$)

# FAQ 62. How Do I Do Linear Regression?

After making a scatter plot (ch. 59) and deciding that the data looks like it would be best described by a straight line, we would like to get an equation of that line.

Since the data is usually noisy (messy) and forms a cloud of points, all of the points will not, in general, fall on the same line. Thus we need to find some way of finding the line that **best describes** the data rather than a line that goes through the data. We do this with the **method of least squares**, also called **least squared regression**. .

Suppose that we draw some line through the data. We then drop vertical lines from each point to the line(62.1). In theory we could measure the lengths of all of these distances, add up the lengths, and jiggle the original line around until the sum is as small as can be.[1]

We write the **linear regression of** $y$ **on** $x$ as an equation

$$\hat{y} = a + bx$$

This is read as **y hat**. The hat notation denotes predicted values. For any given value of $x$, $\hat{y} = a + bx$ is the predicted value of $y$ based on the best fit line.

Figure 62.1.: Basic concept of the least squares regression line. The regression of $y$ on $x$ is line that minimizes the sum of the squared vertical distances from the points to the line.

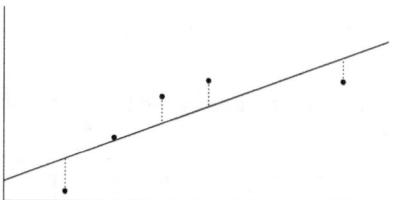

The **slope of the least squares linear regression** of $y$ on $x$ is

$$b = r\frac{s_y}{s_x}$$

where $r$ is the correlation (ch. 61); and $s_x$ and $s_y$ are the calculated standard deviations of the $x$ and $y$ data, respectively (ch. 15).

The **y-intercept of the least squares linear regression** of $y$ on $x$ is

$$a = \bar{y} - b\bar{x}$$

where $\bar{x}$ and $\bar{y}$ are the means of the $x$ and $y$ data respectively; and $b$ is the slope of the line. Sine the y intercept depends on the slope, the slope must be computed first.

The procedure for finding the least squares linear regression is summarized in figure 62.2.

**Example 62.1.** Find the least squares linear regression of y on x for the following

---

[1]For mathematical reasons we add up the sum of the squares of the lengths rather than the lengths themselves to get the results givn here. To derive these equations requires the use of multivariate calculus and so we only present the results.

Figure 62.2.: Procedural flow in calcuating the slope and intercept in the equation for the line of least squares linear regression.

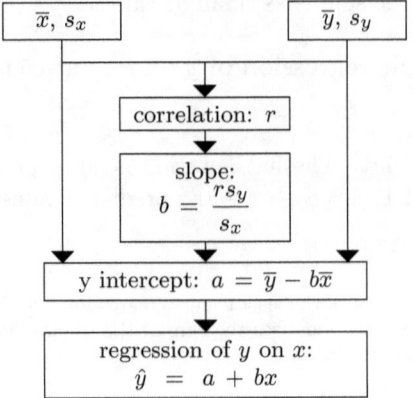

data. This is the same data set that we worked with in example 61.1.

| x | 1.1 | 1.9 | 2. | 2.8 | 3.7 | 4.5 |
|---|-----|-----|-----|------|------|------|
| y | 2.2 | 2.3 | 2.42 | 2.49 | 2.83 | 2.75 |

1. Find $\bar{x}$, the **mean of the explanatory variable data**, i.e., of $x_1, \ldots, x_n$ (ch.13),

$$\bar{x} = \frac{x_1 + x_2 + \cdots + x_n}{n}$$

**(Continuation of example 62.1.)**

$$\bar{x} = \frac{1.1 + 1.9 + 2. + 2.8 + 3.7 + 4.5}{6} = 2.68.$$

2. Find $s_x$, the **standard deviation of the explanatory variable data**, i.e., of $x_1, \ldots, x_n$ (ch. 15), the square root of the variance:

$$s_x^2 = \frac{1}{n-1} \sum_{i=1}^{n} (x_i - \bar{x})$$

where $\bar{x}$ is the **mean of the explanatory variable data** (step 1).

(**Continuation of example 62.1.**)

$$s_x^2 = \frac{1}{n-1} \sum_{i=1}^{n} (x_i - \bar{x})$$

$$= \frac{1}{6-1} \big[ (1.1 - 2.68)^2 + (1.9 - 2.68)^2 + (2 - 2.68)^2$$

$$+ (2.8 - 2.68)^2 + (3.7 - 2.68)^2 + (4.5 - 2.68)^2 \big]$$

$$= 1.59$$

Hence the standard deviation is $s_x = \sqrt{s_x^2} = \sqrt{1.59} = 1.26$

3. Find $\bar{y}$, the **mean of the the response variable data**, i.e., of $y_1, \ldots, y_n$ (ch. 13),

$$\bar{y} = \frac{y_1 + y_2 + \cdots + y_n}{n}$$

(**Continuation of example 62.1.**) As we saw in example 61.1, the average of the y values is

$$\bar{y} = \frac{2.2 + 2.3 + 2.42 + 2.49 + 2.83 + 2.75}{6} = 2.50.$$

4. Find $s_y$, the **standard deviation of the response variable daa**, i.e., of $y_1, \ldots, y_n$ (ch. 15), the square root of the variance:

$$s_y^2 = \frac{1}{n-1} \sum_{i=1}^{n} (y_i - \bar{y})$$

Here $\bar{y}$ is the **mean of the response variable data** (step 3).

(**Continuation of example 62.1.**)

$$s_y^2 = \frac{1}{n-1} \sum_{i=1}^{n} (y_i - \bar{y})$$

$$= \frac{1}{6-1} \big[ (2.2 - 2.5)^2 + (2.3 - 2.5)^2 + (2.42 - 2.5)^2$$

$$+ (2.49 - 2.5)^2 + (2.83 - 2.50)^2 + (2.75 - 2.5)^2 \big]$$

$$= 0.062$$

Thus the standard deviation is $s_y = \sqrt{s_y^2} = \sqrt{0.062} = 0.025$.

---

5. Calculate the correlation $r$ (ch. 61)

$$r = \frac{1}{n-1} \sum_{i=1}^{n} \left[ \frac{x_i - \bar{x}}{s_x} \right] \left[ \frac{y_i - \bar{y}}{s_y} \right]$$

(**Continuation of example 62.1.**) As we also saw in example 61.1, the correlation is

$$= \frac{1}{5} \left\{ \left[ \frac{1.1 - 2.68}{1.26} \right] \left[ \frac{2.2_1 - 2.5}{0.25} \right] + \left[ \frac{1.9 - 2.68}{1.26} \right] \left[ \frac{2.3 - 2.5}{0.25} \right] + \right.$$

$$\left[ \frac{2 - 2.68}{1.26} \right] \left[ \frac{2.42 - 2.5}{0.25} \right] + \left[ \frac{2.8 - 2.68}{1.26} \right] \left[ \frac{2.49 - 2.5}{0.25} \right] +$$

$$\left. \left[ \frac{3.7 - 2.68}{1.26} \right] \left[ \frac{2.83 - 2.5}{0.25} \right] + \left[ \frac{4.5 - 2.68}{1.26} \right] \left[ \frac{2.75 - 2.5}{0.25} \right] \right\}$$

$$= 0.944$$

6. Calculate the slope $b$ using the standard deviations $s_x$ and $s_y$ (from steps 2 and 4), and the correlation $r$ (from step 5) using the formula

$$b = \frac{r s_y}{s_x}$$

(**Continuation of example 62.1.**) Since $r = 0.944$, $s_y = 0.25$ and $s_x = 1.26$,

$$b = \frac{r s_y}{s_x} = \frac{(0.944) \times (0.25)}{1.26} = 0.187$$

7. Calculate the $y$-intercept $a$ using the slope from step 6, along with the sample means $\bar{x}$ and $\bar{y}$ (from steps 1 and 3), using the formula

$$a = \bar{y} - b\bar{x}$$

(**Continuation of example 62.1.**)  Since $\bar{y} = 2.50$, $\bar{x} = 2.68$, and $b = 0.187$, the intercept is

$$a = \bar{y} - b\bar{x} = 2.50 - (0.187) \times (2.68) = 1.997$$

8. The least squares line of linear regression is given by

$$\hat{y} = a + bx$$

where $\hat{y}$ is the **predicted value** of $y$ at $x$, and $a$ and $b$ are the $y$-intercept and slope found in steps 7 and 6.

(**Continuation of example 62.1.**) Since $a = 1.997$ and $b = 0.187$, the regression of $y$ on $x$ is given by

$$\hat{y} = 1.997 + 0.187x$$

**Technology Example 62.2 (TI-84).** Find the slope and $y$ intercept for the regression of $y$ on $x$ for the data given in example 61.1 using a TI-84.

See example 61.2. The calculation of least squares on the TI-82 is identical to the calculation of the correlation because both sets of results are displayed at the same time. The values of $a$ and $b$ can be read off the final screen in the calculation:

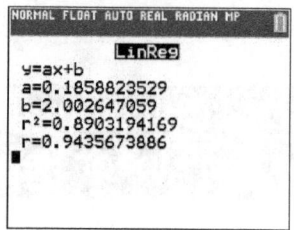

The line of least squares is $y = ax + b$ where $a$ is **0.1858823529** and $b$ is **2.0022657059**.

**Technology Example 62.3 (R).** Repeat example 62.1 using R. Use the function **lm(y~x)** (short for linear model).

```
> x=c(1.1,1.9,2,2.8,3.7,4.5)
> y=c(2.2,2.3,2.42,2.49,2.83,2.75)
> lm(y~x)
Call:
lm(formula = y ~ x)
Coefficients:
(Intercept)          x
    2.0026       0.1859
```

# Exercises

Find the line of least squares linear regression for each of the following data sets. (You found the correlation to each of these data sets in chapter 61.)

1. (23, 19.9), (54, 12.8), (84, 35.0), (28, 9.8), (88, 38.1), (22, 6.5), (11, 1.0), (90, 39.7), (11, 11.1), (64, 26.7), (40, 22.3), (85, 29.2), (12, 6.4), (53, 22.1), (51, 12.7) (ans: $\bar{x} = 47.733$, $\bar{y} = 19.553$, $s_x = 29.463$, $s_y = 11.814$, $r = 0.901$, $a = 2.312$, $b = 0.361$, $y = 9.8$)

2. (90, -68.4), (10, -9.5), (31, -21.4), (43, -27.2), (72, -64.2), (33, -32.1), (72, -46.7), (75, -49.4), (82, -70.5), (57, -49.0), (26, -21.8), (34, -34.4), (61, -52.5), (13, -14.0) (ans: $\bar{x} = 49.929$, $\bar{y} = -40.079$, $s_x = 26.146$, $s_y = 19.335$, $r = -0.955$, $a = -4.818$, $b = -0.706$ )

3. (22, 11.9), (21, 15.0), (39, 8.6), (21, 7.2), (19, 11.2), (58, 29.5), (80, 35.8), (29, 18.5), (40, 25.0), (14, -0.6), (89, 33.9), (78, 22.2), (19, 12.4), (71, 27.0), (26, 17.1) (ans: $\bar{x} = 41.733$, $\bar{y} = 18.313$, $s_x = 26.196$, $s_y = 10.079$, $r = 0.852$, $a = 4.628$, $b = 0.328$)

4. (87, 28.0), (28, 7.3), (26, 14.8), (34, 14.4), (21, 15.5), (89, 36.4), (73, 24.7), (66, 18.7), (47, 20.3), (81, 20.0), (20, 9.1), (33, 15.2), (70, 18.1), (66, 30.8), (89, 32.6) (ans: $\bar{x} = 55.333$, $\bar{y} = 20.393$, $s_x = 26.378$, $s_y = 8.227$, $r = 0.843$, $a = 5.853$, $b = 0.263$)

5. (57, -13.5), (69, -3.1), (50, 21.4), (90, 9.9), (62, 30.2), (137, -18.1), (29, 27.3), (55, 21.5), (32, 24.1), (59, 2.0), (71, 4.8), (29, 13.2), (92, 36.2), (44, 22.6), (140, 0.6), (132, 16.1), (84, -18.2) (ans: $\bar{x} = 72.471$, $\bar{y} = 10.412$, $s_x = 35.945$, $s_y = 16.353$, $r = -0.399$, $a = 23.556$, $b = -0.181$)

---

# FAQ 63. How Do I Calculate the Predicted Value in a Linear Regression?

The **Predicted Value** for a given value of $x$ in a Linear Regression

$$\hat{y} = a + bx$$

is the value of $\hat{y}$ obtained when that value of $x$ is substituted in the equation for the linear regression.

$$\boxed{\text{Predicted Value at x}} = a + b \times \boxed{\text{Value of x}}$$

To find the predicted value (of $y$) for a given value of $x$ using least squares,

1. Find the equation for linear regression of $y$ on $x$, $\hat{y} = a + bx$. If this is given, skip to step 2; otherwise, find $a$ and $b$ as described in chapter 62 first.

**Example 63.1.** In example 62.1 we found that the least squares linear regression for the following data:

| x | 1.1 | 1.9 | 2. | 2.8 | 3.7 | 4.5 |
|---|---|---|---|---|---|---|
| y | 2.2 | 2.3 | 2.42 | 2.49 | 2.83 | 2.75 |

is given by $\hat{y} = 1.997 + 0.187x$. We want to find the predicted value of $y$ at $x = 3$.

2. Substitute the given value of $x$ into the right-hand side of the regression equation.

**Example 63.2.** To find the predicted value at $x = 3$, substitute the value 3 for $x$, i.e., calculate $a + 3b$. The predicted value at $x = 3$ is

$$\hat{y} = 1.997 + 0.187x = 1.997 + 0.187(3) = 2.558$$

The *actual value* at $x = 3$ is *unknown*.

# Exercises

Find the predicted values of the linear regressions for each of the given fits at the specified points.

1. $a = 5$, $b = 2.7$, $x = 4$
2. $a = 10.7$, $b = -3$, $x = 1.5$
3. $a = 7$, $b = -4.2$, $x = 10$
4. $\hat{y} = 3x - 12$, $x = 11$
5. $\hat{y} = 7.6x$, $x = -3.50$
6. $\hat{y} = 3.17 + 12.50x$, $x = 1.43$

# FAQ 64. What is a Residual?

A **residual** is the difference between the actual and predicted value in a linear regression.

$$\boxed{\text{residual}} = \boxed{\text{actual value}} - \boxed{\text{predicted value}}$$

The residual for any particular point $(x, y)$ is the **vertical distance from that point to the line**, where points above the line have positive residuals, and points below the line have negative residuals.

If we have a collection of data values $(x_i, y_i)$, and a least squares prediction $\hat{y} = a + bx$, then the *predicted value* at $x_i$ is $\hat{y}_i = a + bx_i$. Therefore the residual at $x_i$ is

$$\text{residual at } x_i = y_i - \hat{y}_i = y_i - (a + bx_i)$$

To find the residual at $x$:

1. Find the equation for linear regression of $y$ on $x$, $\hat{y} = a + bx$. If this is given, skip to step 2; otherwise, find $a$ and $b$ as described in chapter 62.

**Example 64.1.** Find the residual at $x = 2$ for the data set used in example 62.1. The least squares regression line found there was $\hat{y} = 1.997 + 0.187x$.

2. Calculated the predicted value at $x$:

$$\text{predicted value} = a + bx$$

**(Continuation of example 64.1.)**
At $x = 2$, the regression line $\hat{y} = 1.997 + 0.187x$ gives

$$\text{predicted value} = 1.997 + 0.187(2) = 2.371$$

3. Determine the observed value of $y$ at $x$. This is always given in the original data set. If this is not given then it is not possible to determine the residual at $x$. It is only possible to determine a residual at a point that is given in the original data set.

**(Continuation of example 64.1.)** Looking at the data table in example 62.1, the value of $y$ corresponding to $x = 2$ is $y = 2.42$. This is the observed value:

$$\text{observed value} = 2.42$$

4. Calculate the residual as

$$(\text{residual}) = (\text{observed value}) - (\text{predicted value}) = y - \hat{y}$$

**(Continuation of example 64.1.)** Since the observed value is $y = 2.42$ and the predicted value is $\hat{y} = 2.371$, the residual at $x = 2$ is

$$(\text{residual}) = y - \hat{y} = 2.42 - 2.371 = 0.049$$

# Exercises

Find the predicted values and residuals for each of the following data sets at the specified points. These are same data sets used in the exercises in chapter 62.

1. (23, 19.9), (54, 12.8), (84, 35.0), (28, 9.8), (88, 38.1), (22, 6.5), (11, 1.0), (90, 39.7), (11, 11.1), (64, 26.7), (40, 22.3), (85, 29.2), (12, 6.4), (53, 22.1), (51, 12.7) at $x = 28$ (ans: $\hat{y} = 12.425$, residual $= -2.625$)

2. (90, -68.4), (10, -9.5), (31, -21.4),

(43, -27.2), (72, -64.2), (33, -32.1), (72, -46.7), (75, -49.4), (82, -70.5), (57, -49.0), (26, -21.8), (34, -34.4), (61, -52.5), (13, -14.0) at $x = 43$ (ans: $y = -27.2$, $\hat{y} = -35.0$, residual= 7.986)

3. (22, 11.9), (21, 15.0), (39, 8.6), (21, 7.2), (19, 11.2), (58, 29.5), (80, 35.8), (29, 18.5), (40, 25.0), (14, -0.6), (89, 33.9), (78, 22.2), (19, 12.4), (71, 27.0), (26, 17.1) at $x = 21$ (ans: $\hat{y} = 11.514$, residual $= -4.314$)

4. (87, 28.0), (28, 7.3), (26, 14.8),

(34, 14.4), (21, 15.5), (89, 36.4), (73, 24.7), (66, 18.7), (47, 20.3), (81, 20.0), (20, 9.1), (33, 15.2), (70, 18.1), (66, 30.8), (89, 32.6) at $x = 34$ (ans: $\hat{y} = 15.0$, residual $= -0.387$)

5. (57, -13.5), (69, -3.1), (50, 21.4), (90, 9.9), (62, 30.2), (137, -18.1), (29, 27.3), (55, 21.5), (32, 24.1), (59, 2.0), (71, 4.8), (29, 13.2), (92, 36.2), (44, 22.6), (140, 0.6), (132, 16.1), (84, -18.2) at $x = 90$ (ans: $\hat{y} = 7.0$, residual $= 2.667$)

# FAQ 65. How Do I Install R on My Computer?

Although R includes an interactive command console, it is easier to use it via a graphical user interface (GUI) such as RStudio. Both R and RStudio are free programs that you are allowed to download and install on as many computers as you like at no cost.

## Installation on Mac OS

1. Download the appropriate R binary installer for your operating system.

   The binary can be found at https://cran.r-project.org/bin/macosx/.

   If you have MacOS 10.9 or higher, you probably want the file **R-3.3.2.pkg**.[1] If you have an earlier version of MacOs, scroll further down the page https://cran.r-project.org/bin/macosx/ until you find a version that is compatible with your system.

   If you have a really old Mac and need a 32 bit version you should check the instructions at https://cran.r-project.org/bin/

---

[1] A more recent version may have been released since the publication of this book, and if so, you should download that.

`macosx/RMacOSX-FAQ.html` for further information.

2. Locate the download file, e.g., **r-3.3.2.pkg** in your **downloads** folder. (You can locate the Downloads folder by clicking on the desktop and then from the **Finder** menu select  Go  ⟩⟩ Downloads . On many Macs the key combination  Option + Command +  L  will also take you to the downloads folder. )

3. Double click on the icon for **r-3.3.2.pkg**.

   This should open the installer welcome screen, which will look something like figure 65.1.

Figure 65.1.: The R installer welcome screen.

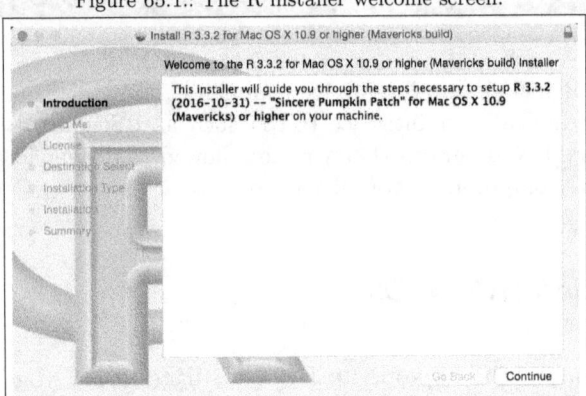

If the install screen does not open, you may have third-part installations disabled (fig. 65.2).

To enable third party installations, go to

System Preferences ⟩⟩ Security and Privacy ⟩⟩ General

On the bottom, under where it says  Allow apps downloaded from , make sure the button  Anywhere  is checked. This part of the screen is normally grayed out; to enable it, click on the little lock on the bottom left of the panel and enter your password. Then click on the lock again after you have clicked on  Anywhere .

Figure 65.2.: The general security control panel.

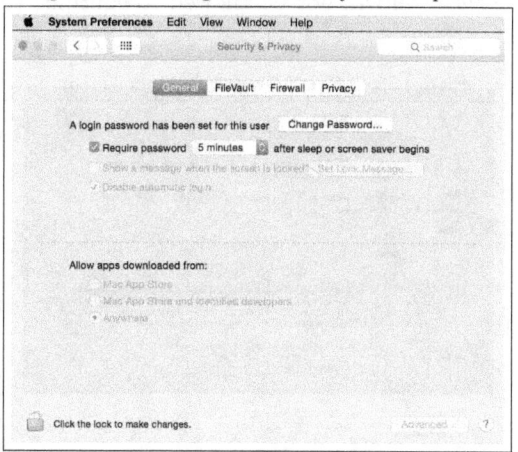

4. On the R install welcome screen, click Continue. A window labeled "important information" will appear. You can read this now, save it as a text file, or print it out and read it later (fig. 65.3).

Figure 65.3.: The installer will provide you with some general information about the contents of the installation.

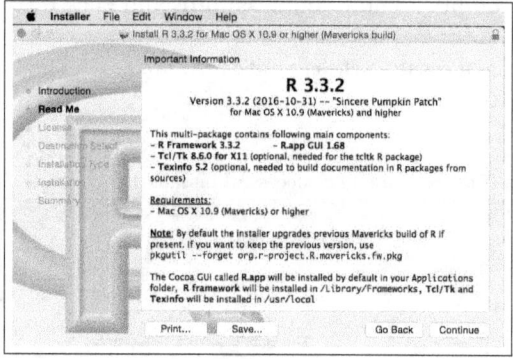

5. After you have read the "important information," click Continue. A window with the software license agreement will appear. Click Continue.

6. A pop-up asking if you agree with the terms of the software license agreement appear. If you want to read the license agreement first,

click on Read License. If you agree, click on Agree. If you do not click on Agree you will not be allowed to continue (fig. 65.4).

Figure 65.4.: The license agreement menu.

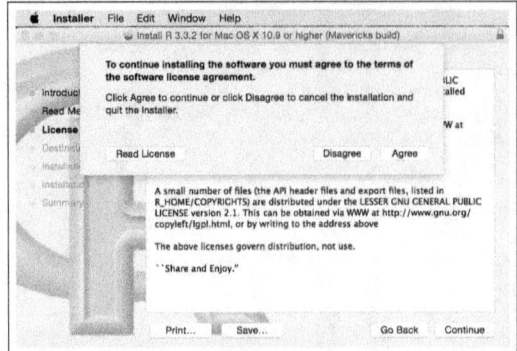

7. You will be asked to select a destination. Most people should select **"Install for all users of this computer."** Click Continue.

8. You will be asked to verify. "Click install to perform a standard installation of this software for all users of this computer ..." You should click Install.

9. When prompted, enter you password, and click on Install Software.

10. When the message The installation was successful (fig. 65.5) appears click on Close.

Figure 65.5.: The successful installation message.

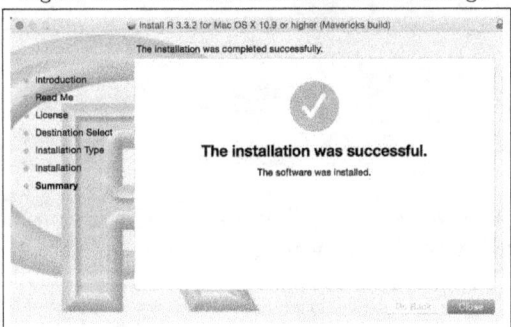

11. Locate the file **R.app** in the applications folder of your computer.

(You can locate the Applications folder by clicking on the desktop and then Finder 》 Go 》 Applications. On some Macs the key combination Shift + Command + A will also take you to the the applications folder.) Double click on R.app it to verify that R has been successfully installed. You should see the R console (fig. 65.6).

Figure 65.6.: The standard R console in MacOS.

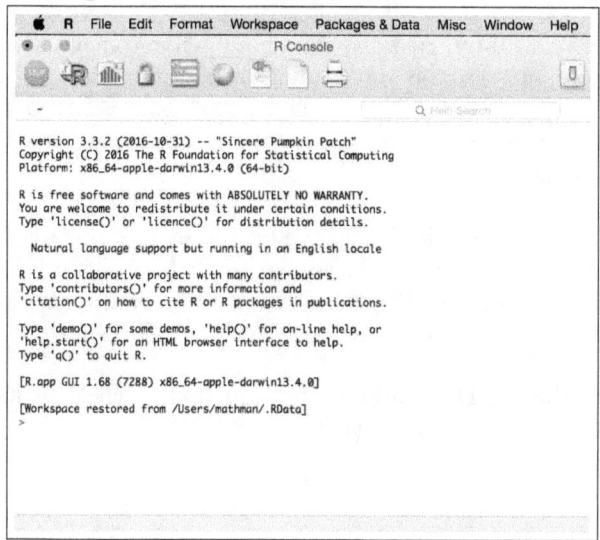

12. Quit from the R Console (e.g., menu bar R 》 Quit R or command + Q ).

13. Download RStudio from https://www.rstudio.com/products/rstudio/download3/

   Make sure you scroll down to the section on free installers. Click on the one for MacOS. At the time this was written it was version 1.044, and a direct link was https://download1.rstudio.org/RStudio-1.0.44.dmg.

14. Locate the download file (e.g., **RStudio-1.0.44.dmg**). It will be in the **downloads** folder.

15. Double click on the **dmg** file icon. This will mount a virtual disk with the same name as the download but without the dmg file extension **RStudio-1.0.44**. When mounted, the contents of the virtual disk

will look like fig. 65.7.

Figure 65.7.: Contents of the R installer virtual disk (dmg file) in MacOS.

16. If a window for virtual disk does not open automatically, locate the disk on your desktop and double click on it.

17. Drag the **RStudio.app** icon onto the **Applications** folder icon (fig. 65.7).

18. If you have a standard Mac one-button mouse, control click on the RStudio virtual disk and select eject. If have a two button mouse right click and select eject. With a track-pad, do a two fingered click and select eject.

19. Open the **Applications** folder and double click on **RStudio** to verify it has been installed (fig. 65.8).

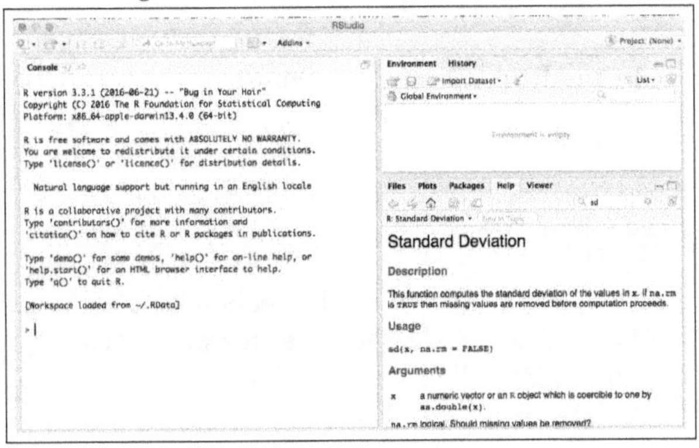

Figure 65.8.: RStudio console in MacOS.

## Installation in Windows

1. Download the latest version of the R installer for Windows from https://cran.r-project.org/bin/windows/base/.

At the time this was written the latest version was called **R-3.3.2 for Windows (32/64 bit)**. This version should run on most versions of Windows XP, Vista, 7, 8 and 10, both 32 and 64 bit versions.

2. Locate the installer file after the download has completed. It should be in your **downloads** folder. This is normally the folder C ⟩ Users ⟩ ⟩ your-user-name ⟩ Downloads. You should also be able to get to this folder in Windows 10 by entering Downloads into the search bar. The file name is **R-3.3.2-win.exe**.

3. A pop up will ask if you want to allow an unknown publisher to make changes to your device (fig. 65.9). You should click Yes.

Figure 65.9.: Click yes to allow installation of new software in Windows 10.

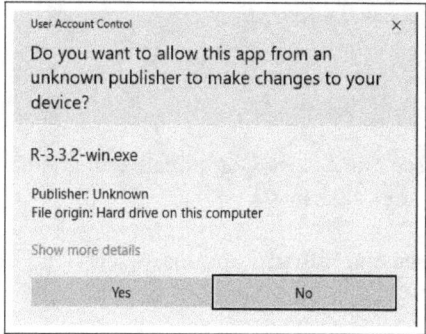

4. A pop up will ask for the install language, with the default set to English (fig. 65.10). Click ok.

Figure 65.10.: Select the language your want to use. English is the default..

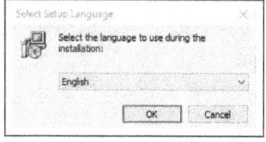

5. When you see the "Welcome to the R for Windows Setup Wizard" click next (fig 65.11).

6. When you see the Software License click next, accepting the license.

7. When you see the Software Destination click next, accepting the

Figure 65.11.: The Windows install wizeard begins.

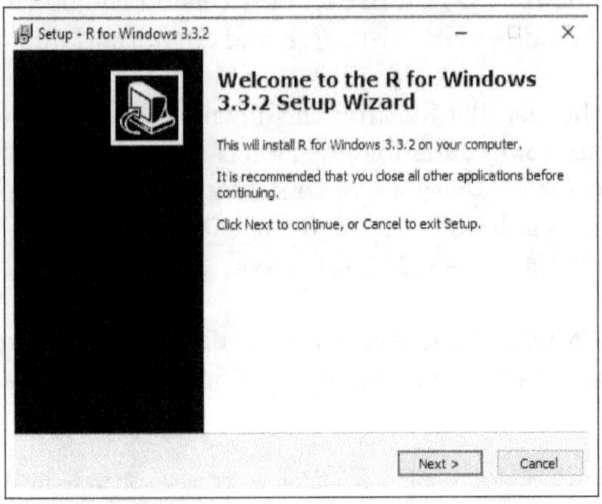

default install location.

8. When you see the Select Components, make sure all components are checked and click next.

9. When you see the Startup options, select "no" to accept defaults, and click next.

10. When you see the Select Start Menu Folder, do not make any changes, accept the default settings to include R in the start menu, and click next.

11. When you see Select Additional Tasks, make sure that the two boxes under additional icons are not checked (one of them is normally checked by default, so you will have to uncheck). You should not create a desktop icon for R because you will be opening it through RStudio and not directly. Leave the Registry check boxes in their default settings. Click next.

12. A progress bar will be displayed while the software is being installed. When the installation is completed a message will be displayed that says "Click Finish to exit Setup" (fig. 65.12). You should do so.

13. If you want to verify that the application was installed, locate the R console application and double click on it to make sure it loads.

Figure 65.12.: The Windows installation of R is complete.

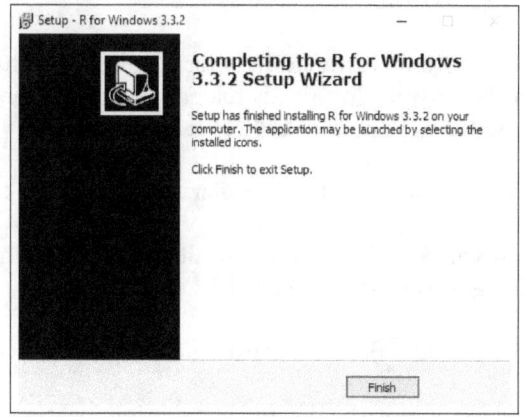

The application is called **R.exe** and is located in the folder [C] [Program Files] [R] [R-3.3.2] [bin]. It should look something like fig. 65.13. Quit from the R console by clicking in the close box (the x at the top right corner).

Figure 65.13.: The R Console in Windows 10.

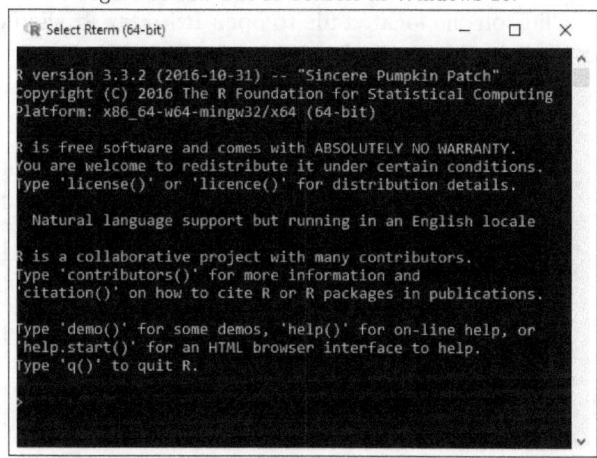

14. Download RStudio for Windows from https://www.rstudio.com/products/rstudio/download3/. Locate the free version that applies to your operating system. At the time that this was written, the most recent version was RStudio-1.0.44.exe.

A direct link to the download file was https://download1.
rstudio.org/RStudio-1.0.44.exe.

15. Locate the RStudio installer file after the download has completed.
It should be in your downloads folder. Look for a file with a name
that is similar to **RStudio-1.0.44.exe**.

16. Double click on icon for the installer **RStudio-1.0.44.exe**.

17. A pop up will ask if you want to allow an unknown publisher to
make changes to your device. Click Yes .

18. When the RStudio Setup Wizard menu opens, click next .

19. Click next to accept the default installation location.

20. Click next to accept the default start menu folder.

21. A progress bar will be displayed while the installation progresses.
Click finish when it says it is complete.

22. To verify that RStudio has been properly installed, type RStudio
into the Windows search bar. It should locate it as a "Desktop
app." Click on the located file to open RStudio. It should look like
fig 65.14.

Figure 65.14.: RStudio in Windows 10.

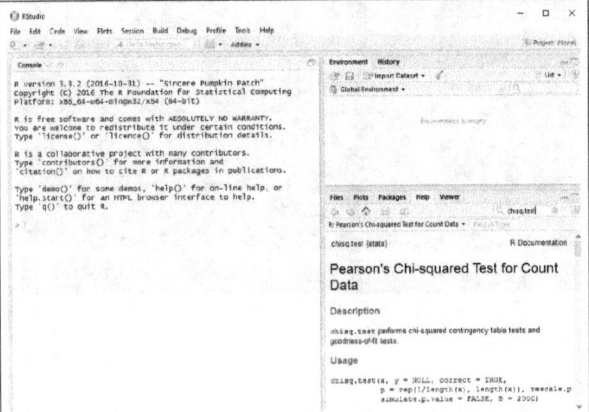

# FAQ 66. How Does My Graphing or Statistical Calculator Work?

There are bazillions of teeny-tiny little magic Fairies inside of it, and every time you push a button one flies off to Neverland while flapping his wings to the tune of *Beat It*. "You better run, you better do what you can." Once there, the Fairy rubs a magic lamp. The lamp connects the Fairy directly to the Far-Far-Away IT department genie. This genie, whose name is Gene, is somewhat temperamental. Sometimes Gene gives the Fairy erroneous information. Gene says, "You rubbed my lamp. Just beat it. You want to be tough, better do what you can

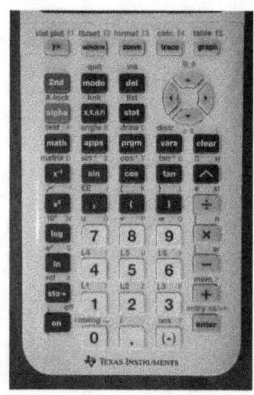

So beat it." The fairy from behind the buttons thinks, "I want to be bad, want to do what I can," so he beats the lamp. This puts Gene in a more magnanimous mood. He figures out the answer and gives it to the Fairy, saying "You better run, you better do what you can. Don't want to see no blood. Don't be a macho man." So Fairy number one flies back to your calculator post-haste and pushes teeny-tiny buttons behind the display to light up all the right pixels on your display. This happens really fast, faster than you could hop to the top of the Empire State Building backwards on one foot with your hands tied behind your back. Then the Fairy eats a light meal, maybe a pot roast or lasagna, and goes to sleep.

# FAQ 67. Have You Ever Dated Another Statistician?

Seriously? In the first place, I am not a statistician. And in the second place, ewww.

# FAQ 68. Will All This Be on the Final?

Duh. Yes.

# Appendix A. Statistical Tables

Table A.1 Cumulative distribution of the standard normal distribution. The table entries give the area to under a standard normal distribution to the left of normalized z-value as described in chapter 21.

Table A.2 Critical values of the t-distribution for various degrees of freedom. See chapter 39.

Table A.3 Critical values of the chi-squared distribution. See chapter 56.

Table A.1.: Cumulative distribution for the standard normal distribution for $z < 0$

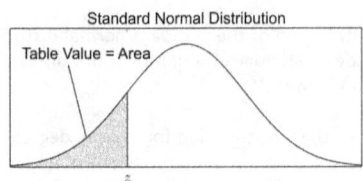

| | 0 | 0.01 | 0.02 | 0.03 | 0.04 | 0.05 | 0.06 | 0.07 | 0.08 | 0.09 |
|------|--------|--------|--------|--------|--------|--------|--------|--------|--------|--------|
| -3.4 | 0.0003 | 0.0003 | 0.0003 | 0.0003 | 0.0003 | 0.0003 | 0.0003 | 0.0003 | 0.0003 | 0.0002 |
| -3.3 | 0.0005 | 0.0005 | 0.0005 | 0.0004 | 0.0004 | 0.0004 | 0.0004 | 0.0004 | 0.0004 | 0.0003 |
| -3.2 | 0.0007 | 0.0007 | 0.0006 | 0.0006 | 0.0006 | 0.0006 | 0.0006 | 0.0005 | 0.0005 | 0.0005 |
| -3.1 | 0.0010 | 0.0009 | 0.0009 | 0.0009 | 0.0008 | 0.0008 | 0.0008 | 0.0008 | 0.0007 | 0.0007 |
| -3   | 0.0013 | 0.0013 | 0.0013 | 0.0012 | 0.0012 | 0.0011 | 0.0011 | 0.0011 | 0.0010 | 0.0010 |
| -2.9 | 0.0019 | 0.0018 | 0.0018 | 0.0017 | 0.0016 | 0.0016 | 0.0015 | 0.0015 | 0.0014 | 0.0014 |
| -2.8 | 0.0026 | 0.0025 | 0.0024 | 0.0023 | 0.0023 | 0.0022 | 0.0021 | 0.0021 | 0.0020 | 0.0019 |
| -2.7 | 0.0035 | 0.0034 | 0.0033 | 0.0032 | 0.0031 | 0.0030 | 0.0029 | 0.0028 | 0.0027 | 0.0026 |
| -2.6 | 0.0047 | 0.0045 | 0.0044 | 0.0043 | 0.0041 | 0.0040 | 0.0039 | 0.0038 | 0.0037 | 0.0036 |
| -2.5 | 0.0062 | 0.0060 | 0.0059 | 0.0057 | 0.0055 | 0.0054 | 0.0052 | 0.0051 | 0.0049 | 0.0048 |
| -2.4 | 0.0082 | 0.0080 | 0.0078 | 0.0075 | 0.0073 | 0.0071 | 0.0069 | 0.0068 | 0.0066 | 0.0064 |
| -2.3 | 0.0107 | 0.0104 | 0.0102 | 0.0099 | 0.0096 | 0.0094 | 0.0091 | 0.0089 | 0.0087 | 0.0084 |
| -2.2 | 0.0139 | 0.0136 | 0.0132 | 0.0129 | 0.0125 | 0.0122 | 0.0119 | 0.0116 | 0.0113 | 0.0110 |
| -2.1 | 0.0179 | 0.0174 | 0.0170 | 0.0166 | 0.0162 | 0.0158 | 0.0154 | 0.0150 | 0.0146 | 0.0143 |
| -2   | 0.0228 | 0.0222 | 0.0217 | 0.0212 | 0.0207 | 0.0202 | 0.0197 | 0.0192 | 0.0188 | 0.0183 |
| -1.9 | 0.0287 | 0.0281 | 0.0274 | 0.0268 | 0.0262 | 0.0256 | 0.0250 | 0.0244 | 0.0239 | 0.0233 |
| -1.8 | 0.0359 | 0.0351 | 0.0344 | 0.0336 | 0.0329 | 0.0322 | 0.0314 | 0.0307 | 0.0301 | 0.0294 |
| -1.7 | 0.0446 | 0.0436 | 0.0427 | 0.0418 | 0.0409 | 0.0401 | 0.0392 | 0.0384 | 0.0375 | 0.0367 |
| -1.6 | 0.0548 | 0.0537 | 0.0526 | 0.0516 | 0.0505 | 0.0495 | 0.0485 | 0.0475 | 0.0465 | 0.0455 |
| -1.5 | 0.0668 | 0.0655 | 0.0643 | 0.0630 | 0.0618 | 0.0606 | 0.0594 | 0.0582 | 0.0571 | 0.0559 |
| -1.4 | 0.0808 | 0.0793 | 0.0778 | 0.0764 | 0.0749 | 0.0735 | 0.0721 | 0.0708 | 0.0694 | 0.0681 |
| -1.3 | 0.0968 | 0.0951 | 0.0934 | 0.0918 | 0.0901 | 0.0885 | 0.0869 | 0.0853 | 0.0838 | 0.0823 |
| -1.2 | 0.1151 | 0.1131 | 0.1112 | 0.1093 | 0.1075 | 0.1056 | 0.1038 | 0.1020 | 0.1003 | 0.0985 |
| -1.1 | 0.1357 | 0.1335 | 0.1314 | 0.1292 | 0.1271 | 0.1251 | 0.1230 | 0.1210 | 0.1190 | 0.1170 |
| -1   | 0.1587 | 0.1562 | 0.1539 | 0.1515 | 0.1492 | 0.1469 | 0.1446 | 0.1423 | 0.1401 | 0.1379 |
| -0.9 | 0.1841 | 0.1814 | 0.1788 | 0.1762 | 0.1736 | 0.1711 | 0.1685 | 0.1660 | 0.1635 | 0.1611 |
| -0.8 | 0.2119 | 0.2090 | 0.2061 | 0.2033 | 0.2005 | 0.1977 | 0.1949 | 0.1922 | 0.1894 | 0.1867 |
| -0.7 | 0.2420 | 0.2389 | 0.2358 | 0.2327 | 0.2296 | 0.2266 | 0.2236 | 0.2206 | 0.2177 | 0.2148 |
| -0.6 | 0.2743 | 0.2709 | 0.2676 | 0.2643 | 0.2611 | 0.2578 | 0.2546 | 0.2514 | 0.2483 | 0.2451 |
| -0.5 | 0.3085 | 0.3050 | 0.3015 | 0.2981 | 0.2946 | 0.2912 | 0.2877 | 0.2843 | 0.2810 | 0.2776 |
| -0.4 | 0.3446 | 0.3409 | 0.3372 | 0.3336 | 0.3300 | 0.3264 | 0.3228 | 0.3192 | 0.3156 | 0.3121 |
| -0.3 | 0.3821 | 0.3783 | 0.3745 | 0.3707 | 0.3669 | 0.3632 | 0.3594 | 0.3557 | 0.3520 | 0.3483 |
| -0.2 | 0.4207 | 0.4168 | 0.4129 | 0.4090 | 0.4052 | 0.4013 | 0.3974 | 0.3936 | 0.3897 | 0.3859 |
| -0.1 | 0.4602 | 0.4562 | 0.4522 | 0.4483 | 0.4443 | 0.4404 | 0.4364 | 0.4325 | 0.4286 | 0.4247 |
| -0.0 | 0.5000 | 0.4960 | 0.4920 | 0.4880 | 0.4840 | 0.4801 | 0.4761 | 0.4721 | 0.4681 | 0.4641 |

Appendix A. Statistical Tables

Table A.1.: Cumulative distribution for the standard normal distribution (continue). Contains values for $z \geqslant 0$

|     | 0      | 0.01   | 0.02   | 0.03   | 0.04   | 0.05   | 0.06   | 0.07   | 0.08   | 0.09   |
|-----|--------|--------|--------|--------|--------|--------|--------|--------|--------|--------|
| 0   | 0.5000 | 0.5040 | 0.5080 | 0.5120 | 0.5160 | 0.5199 | 0.5239 | 0.5279 | 0.5319 | 0.5359 |
| 0.1 | 0.5398 | 0.5438 | 0.5478 | 0.5517 | 0.5557 | 0.5596 | 0.5636 | 0.5675 | 0.5714 | 0.5753 |
| 0.2 | 0.5793 | 0.5832 | 0.5871 | 0.5910 | 0.5948 | 0.5987 | 0.6026 | 0.6064 | 0.6103 | 0.6141 |
| 0.3 | 0.6179 | 0.6217 | 0.6255 | 0.6293 | 0.6331 | 0.6368 | 0.6406 | 0.6443 | 0.6480 | 0.6517 |
| 0.4 | 0.6554 | 0.6591 | 0.6628 | 0.6664 | 0.6700 | 0.6736 | 0.6772 | 0.6808 | 0.6844 | 0.6879 |
| 0.5 | 0.6915 | 0.6950 | 0.6985 | 0.7019 | 0.7054 | 0.7088 | 0.7123 | 0.7157 | 0.7190 | 0.7224 |
| 0.6 | 0.7257 | 0.7291 | 0.7324 | 0.7357 | 0.7389 | 0.7422 | 0.7454 | 0.7486 | 0.7517 | 0.7549 |
| 0.7 | 0.7580 | 0.7611 | 0.7642 | 0.7673 | 0.7704 | 0.7734 | 0.7764 | 0.7794 | 0.7823 | 0.7852 |
| 0.8 | 0.7881 | 0.7910 | 0.7939 | 0.7967 | 0.7995 | 0.8023 | 0.8051 | 0.8078 | 0.8106 | 0.8133 |
| 0.9 | 0.8159 | 0.8186 | 0.8212 | 0.8238 | 0.8264 | 0.8289 | 0.8315 | 0.8340 | 0.8365 | 0.8389 |
| 1   | 0.8413 | 0.8438 | 0.8461 | 0.8485 | 0.8508 | 0.8531 | 0.8554 | 0.8577 | 0.8599 | 0.8621 |
| 1.1 | 0.8643 | 0.8665 | 0.8686 | 0.8708 | 0.8729 | 0.8749 | 0.8770 | 0.8790 | 0.8810 | 0.8830 |
| 1.2 | 0.8849 | 0.8869 | 0.8888 | 0.8907 | 0.8925 | 0.8944 | 0.8962 | 0.8980 | 0.8997 | 0.9015 |
| 1.3 | 0.9032 | 0.9049 | 0.9066 | 0.9082 | 0.9099 | 0.9115 | 0.9131 | 0.9147 | 0.9162 | 0.9177 |
| 1.4 | 0.9192 | 0.9207 | 0.9222 | 0.9236 | 0.9251 | 0.9265 | 0.9279 | 0.9292 | 0.9306 | 0.9319 |
| 1.5 | 0.9332 | 0.9345 | 0.9357 | 0.9370 | 0.9382 | 0.9394 | 0.9406 | 0.9418 | 0.9429 | 0.9441 |
| 1.6 | 0.9452 | 0.9463 | 0.9474 | 0.9484 | 0.9495 | 0.9505 | 0.9515 | 0.9525 | 0.9535 | 0.9545 |
| 1.7 | 0.9554 | 0.9564 | 0.9573 | 0.9582 | 0.9591 | 0.9599 | 0.9608 | 0.9616 | 0.9625 | 0.9633 |
| 1.8 | 0.9641 | 0.9649 | 0.9656 | 0.9664 | 0.9671 | 0.9678 | 0.9686 | 0.9693 | 0.9699 | 0.9706 |
| 1.9 | 0.9713 | 0.9719 | 0.9726 | 0.9732 | 0.9738 | 0.9744 | 0.9750 | 0.9756 | 0.9761 | 0.9767 |
| 2   | 0.9772 | 0.9778 | 0.9783 | 0.9788 | 0.9793 | 0.9798 | 0.9803 | 0.9808 | 0.9812 | 0.9817 |
| 2.1 | 0.9821 | 0.9826 | 0.9830 | 0.9834 | 0.9838 | 0.9842 | 0.9846 | 0.9850 | 0.9854 | 0.9857 |
| 2.2 | 0.9861 | 0.9864 | 0.9868 | 0.9871 | 0.9875 | 0.9878 | 0.9881 | 0.9884 | 0.9887 | 0.9890 |
| 2.3 | 0.9893 | 0.9896 | 0.9898 | 0.9901 | 0.9904 | 0.9906 | 0.9909 | 0.9911 | 0.9913 | 0.9916 |
| 2.4 | 0.9918 | 0.9920 | 0.9922 | 0.9925 | 0.9927 | 0.9929 | 0.9931 | 0.9932 | 0.9934 | 0.9936 |
| 2.5 | 0.9938 | 0.9940 | 0.9941 | 0.9943 | 0.9945 | 0.9946 | 0.9948 | 0.9949 | 0.9951 | 0.9952 |
| 2.6 | 0.9953 | 0.9955 | 0.9956 | 0.9957 | 0.9959 | 0.9960 | 0.9961 | 0.9962 | 0.9963 | 0.9964 |
| 2.7 | 0.9965 | 0.9966 | 0.9967 | 0.9968 | 0.9969 | 0.9970 | 0.9971 | 0.9972 | 0.9973 | 0.9974 |
| 2.8 | 0.9974 | 0.9975 | 0.9976 | 0.9977 | 0.9977 | 0.9978 | 0.9979 | 0.9979 | 0.9980 | 0.9981 |
| 2.9 | 0.9981 | 0.9982 | 0.9982 | 0.9983 | 0.9984 | 0.9984 | 0.9985 | 0.9985 | 0.9986 | 0.9986 |
| 3   | 0.9987 | 0.9987 | 0.9987 | 0.9988 | 0.9988 | 0.9989 | 0.9989 | 0.9989 | 0.9990 | 0.9990 |
| 3.1 | 0.9990 | 0.9991 | 0.9991 | 0.9991 | 0.9992 | 0.9992 | 0.9992 | 0.9992 | 0.9993 | 0.9993 |
| 3.2 | 0.9993 | 0.9993 | 0.9994 | 0.9994 | 0.9994 | 0.9994 | 0.9994 | 0.9995 | 0.9995 | 0.9995 |
| 3.3 | 0.9995 | 0.9995 | 0.9995 | 0.9996 | 0.9996 | 0.9996 | 0.9996 | 0.9996 | 0.9996 | 0.9997 |
| 3.4 | 0.9997 | 0.9997 | 0.9997 | 0.9997 | 0.9997 | 0.9997 | 0.9997 | 0.9997 | 0.9997 | 0.9998 |

## Table A.2.: Critical values of the $t$ distribution.

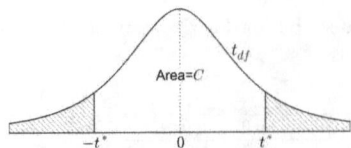

| df | \multicolumn{10}{c}{Confidence Level C} |
|---|---|---|---|---|---|---|---|---|---|---|
| | 50% | 60% | 70% | 80% | 90% | 95% | 96% | 98% | 99% | 99.8% |
| 1 | 1 | 1.376 | 1.963 | 3.078 | 6.314 | 12.706 | 15.895 | 31.821 | 63.657 | 318.309 |
| 2 | 0.816 | 1.061 | 1.386 | 1.886 | 2.92 | 4.303 | 4.849 | 6.965 | 9.925 | 22.327 |
| 3 | 0.765 | 0.978 | 1.25 | 1.638 | 2.353 | 3.182 | 3.482 | 4.541 | 5.841 | 10.215 |
| 4 | 0.741 | 0.941 | 1.19 | 1.533 | 2.132 | 2.776 | 2.999 | 3.747 | 4.604 | 7.173 |
| 5 | 0.727 | 0.92 | 1.156 | 1.476 | 2.015 | 2.571 | 2.757 | 3.365 | 4.032 | 5.893 |
| 6 | 0.718 | 0.906 | 1.134 | 1.44 | 1.943 | 2.447 | 2.612 | 3.143 | 3.707 | 5.208 |
| 7 | 0.711 | 0.896 | 1.119 | 1.415 | 1.895 | 2.365 | 2.517 | 2.998 | 3.499 | 4.785 |
| 8 | 0.706 | 0.889 | 1.108 | 1.397 | 1.86 | 2.306 | 2.449 | 2.896 | 3.355 | 4.501 |
| 9 | 0.703 | 0.883 | 1.1 | 1.383 | 1.833 | 2.262 | 2.398 | 2.821 | 3.25 | 4.297 |
| 10 | 0.7 | 0.879 | 1.093 | 1.372 | 1.812 | 2.228 | 2.359 | 2.764 | 3.169 | 4.144 |
| 11 | 0.697 | 0.876 | 1.088 | 1.363 | 1.796 | 2.201 | 2.328 | 2.718 | 3.106 | 4.025 |
| 12 | 0.695 | 0.873 | 1.083 | 1.356 | 1.782 | 2.179 | 2.303 | 2.681 | 3.055 | 3.93 |
| 13 | 0.694 | 0.87 | 1.079 | 1.35 | 1.771 | 2.16 | 2.282 | 2.65 | 3.012 | 3.852 |
| 14 | 0.692 | 0.868 | 1.076 | 1.345 | 1.761 | 2.145 | 2.264 | 2.624 | 2.977 | 3.787 |
| 15 | 0.691 | 0.866 | 1.074 | 1.341 | 1.753 | 2.131 | 2.249 | 2.602 | 2.947 | 3.733 |
| 16 | 0.69 | 0.865 | 1.071 | 1.337 | 1.746 | 2.12 | 2.235 | 2.583 | 2.921 | 3.686 |
| 17 | 0.689 | 0.863 | 1.069 | 1.333 | 1.74 | 2.11 | 2.224 | 2.567 | 2.898 | 3.646 |
| 18 | 0.688 | 0.862 | 1.067 | 1.33 | 1.734 | 2.101 | 2.214 | 2.552 | 2.878 | 3.61 |
| 19 | 0.688 | 0.861 | 1.066 | 1.328 | 1.729 | 2.093 | 2.205 | 2.539 | 2.861 | 3.579 |
| 20 | 0.687 | 0.86 | 1.064 | 1.325 | 1.725 | 2.086 | 2.197 | 2.528 | 2.845 | 3.552 |
| 21 | 0.686 | 0.859 | 1.063 | 1.323 | 1.721 | 2.08 | 2.189 | 2.518 | 2.831 | 3.527 |
| 22 | 0.686 | 0.858 | 1.061 | 1.321 | 1.717 | 2.074 | 2.183 | 2.508 | 2.819 | 3.505 |
| 23 | 0.685 | 0.858 | 1.06 | 1.319 | 1.714 | 2.069 | 2.177 | 2.5 | 2.807 | 3.485 |
| 24 | 0.685 | 0.857 | 1.059 | 1.318 | 1.711 | 2.064 | 2.172 | 2.492 | 2.797 | 3.467 |
| 25 | 0.684 | 0.856 | 1.058 | 1.316 | 1.708 | 2.06 | 2.167 | 2.485 | 2.787 | 3.45 |
| 30 | 0.683 | 0.854 | 1.055 | 1.31 | 1.697 | 2.042 | 2.147 | 2.457 | 2.75 | 3.385 |
| 40 | 0.681 | 0.851 | 1.05 | 1.303 | 1.684 | 2.021 | 2.123 | 2.423 | 2.704 | 3.307 |
| 50 | 0.679 | 0.849 | 1.047 | 1.299 | 1.676 | 2.009 | 2.109 | 2.403 | 2.678 | 3.261 |
| 60 | 0.679 | 0.848 | 1.045 | 1.296 | 1.671 | 2 | 2.099 | 2.39 | 2.66 | 3.232 |
| 80 | 0.678 | 0.846 | 1.043 | 1.292 | 1.664 | 1.99 | 2.088 | 2.374 | 2.639 | 3.195 |
| 100 | 0.677 | 0.845 | 1.042 | 1.29 | 1.66 | 1.984 | 2.081 | 2.364 | 2.626 | 3.174 |
| 1000 | 0.675 | 0.842 | 1.037 | 1.282 | 1.646 | 1.962 | 2.056 | 2.33 | 2.581 | 3.098 |
| $z^*$ | 0.674 | 0.842 | 1.036 | 1.282 | 1.645 | 1.960 | 2.054 | 2.326 | 2.576 | 3.090 |
| 1-Sided P | 0.25 | 0.2 | 0.15 | 0.1 | 0.05 | 0.025 | 0.02 | 0.01 | 0.005 | 0.001 |
| 2-Sided P | 0.5 | 0.4 | 0.3 | 0.2 | 0.1 | 0.05 | 0.04 | 0.02 | 0.01 | 0.002 |

Appendix A. Statistical Tables

Table A.3.: Chi Squared table of critical values

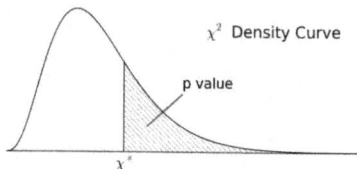

$\chi^2$ Density Curve

p value

$\chi^2$

| df | 0.25 | 0.20 | 0.10 | 0.05 | 0.025 | 0.02 | 0.01 | 0.005 | 0.0025 | 0.001 |
|---|---|---|---|---|---|---|---|---|---|---|
| 1 | 1.32 | 1.64 | 2.71 | 3.84 | 5.02 | 5.41 | 6.63 | 7.88 | 9.14 | 10.83 |
| 2 | 2.77 | 3.22 | 4.61 | 5.99 | 7.38 | 7.82 | 9.21 | 10.60 | 11.98 | 13.82 |
| 3 | 4.11 | 4.64 | 6.25 | 7.81 | 9.35 | 9.84 | 11.34 | 12.84 | 14.32 | 16.27 |
| 4 | 5.39 | 5.99 | 7.78 | 9.49 | 11.14 | 11.67 | 13.28 | 14.86 | 16.42 | 18.47 |
| 5 | 6.63 | 7.29 | 9.24 | 11.07 | 12.83 | 13.39 | 15.09 | 16.75 | 18.39 | 20.52 |
| 6 | 7.84 | 8.56 | 10.64 | 12.59 | 14.45 | 15.03 | 16.81 | 18.55 | 20.25 | 22.46 |
| 7 | 9.04 | 9.80 | 12.02 | 14.07 | 16.01 | 16.62 | 18.48 | 20.28 | 22.04 | 24.32 |
| 8 | 10.22 | 11.03 | 13.36 | 15.51 | 17.53 | 18.17 | 20.09 | 21.95 | 23.77 | 26.12 |
| 9 | 11.39 | 12.24 | 14.68 | 16.92 | 19.02 | 19.68 | 21.67 | 23.59 | 25.46 | 27.88 |
| 10 | 12.55 | 13.44 | 15.99 | 18.31 | 20.48 | 21.16 | 23.21 | 25.19 | 27.11 | 29.59 |
| 11 | 13.70 | 14.63 | 17.28 | 19.68 | 21.92 | 22.62 | 24.72 | 26.76 | 28.73 | 31.26 |
| 12 | 14.85 | 15.81 | 18.55 | 21.03 | 23.34 | 24.05 | 26.22 | 28.30 | 30.32 | 32.91 |
| 13 | 15.98 | 16.98 | 19.81 | 22.36 | 24.74 | 25.47 | 27.69 | 29.82 | 31.88 | 34.53 |
| 14 | 17.12 | 18.15 | 21.06 | 23.68 | 26.12 | 26.87 | 29.14 | 31.32 | 33.43 | 36.12 |
| 15 | 18.25 | 19.31 | 22.31 | 25.00 | 27.49 | 28.26 | 30.58 | 32.80 | 34.95 | 37.70 |
| 16 | 19.37 | 20.47 | 23.54 | 26.30 | 28.85 | 29.63 | 32.00 | 34.27 | 36.46 | 39.25 |
| 17 | 20.49 | 21.61 | 24.77 | 27.59 | 30.19 | 31.00 | 33.41 | 35.72 | 37.95 | 40.79 |
| 18 | 21.60 | 22.76 | 25.99 | 28.87 | 31.53 | 32.35 | 34.81 | 37.16 | 39.42 | 42.31 |
| 19 | 22.72 | 23.90 | 27.20 | 32.85 | 32.85 | 33.69 | 36.19 | 38.58 | 40.88 | 43.82 |
| 20 | 23.83 | 25.04 | 28.41 | 31.41 | 34.17 | 35.02 | 37.57 | 40.00 | 42.34 | 45.31 |
| 21 | 24.93 | 26.17 | 29.62 | 32.67 | 35.48 | 36.34 | 38.93 | 41.40 | 43.78 | 46.80 |
| 22 | 26.04 | 27.30 | 30.81 | 33.92 | 36.78 | 37.66 | 40.29 | 42.80 | 45.20 | 48.27 |
| 23 | 27.14 | 28.43 | 32.01 | 35.17 | 38.08 | 38.97 | 41.64 | 44.18 | 46.62 | 49.73 |
| 24 | 28.24 | 29.55 | 33.20 | 36.42 | 39.36 | 40.27 | 42.98 | 45.56 | 48.03 | 51.18 |
| 25 | 29.34 | 30.68 | 34.38 | 37.65 | 40.65 | 41.57 | 44.31 | 46.93 | 49.44 | 52.62 |
| 30 | 34.80 | 36.25 | 40.26 | 43.77 | 46.98 | 47.96 | 50.89 | 53.67 | 56.33 | 59.70 |
| 40 | 45.62 | 47.27 | 51.81 | 55.76 | 59.34 | 60.44 | 63.69 | 66.77 | 69.70 | 73.40 |
| 50 | 56.33 | 58.16 | 63.17 | 67.50 | 71.42 | 72.61 | 76.15 | 79.49 | 82.66 | 86.66 |
| 60 | 66.98 | 68.97 | 74.40 | 79.08 | 83.30 | 84.58 | 88.38 | 91.95 | 95.34 | 99.61 |
| 80 | 88.13 | 90.41 | 96.58 | 101.88 | 106.63 | 108.07 | 112.33 | 116.32 | 120.10 | 124.84 |
| 100 | 109.14 | 111.67 | 118.50 | 124.34 | 129.56 | 131.14 | 135.81 | 140.17 | 144.29 | 149.45 |

| Parameter | Standard Error Formula | Confidence Interval | Statistic for Hypothesis Test |
|---|---|---|---|
| $\mu$ Population Mean | If you know $\sigma$: $$(\text{s.e.}) = \frac{\sigma}{\sqrt{n}}$$ If you don't know $\sigma$: $$(\text{s.e.}) = \frac{s}{\sqrt{n}}$$ | $$\mu = \bar{x} \pm z^* \times (\text{s.e.}) = \bar{x} \pm z^* \frac{\sigma}{\sqrt{n}}$$ $$\mu = \bar{x} \pm t^* \times (\text{s.e.}) = \bar{x} \pm t^* \frac{s}{\sqrt{n}}$$ | $H_0 : \mu = \mu_0,$ $$z = \frac{\bar{x} - \mu_0}{\sigma/\sqrt{n}}$$ $H_0 : \mu = \mu_0,$ $$t = \frac{\bar{x} - \mu_0}{s/\sqrt{n}}$$ |
| $\mu_1 - \mu_2$ Difference of two means | $$(\text{s.e.}) = \sqrt{\frac{s_1^2}{n_1} + \frac{s_2^2}{n_2}}$$ $$= \sqrt{(\text{s.e.})_1^2 + (\text{s.e.})_2^2}$$ | $$\mu_1 - \mu_2 = \bar{x}_1 - \bar{x}_2 \pm t^* \times (\text{s.e.})$$ $$= \bar{x}_1 - \bar{x}_2 \pm t^* \sqrt{\frac{s_1^2}{n_1} + \frac{s_2^2}{n_2}}$$ $$= \bar{x}_1 - \bar{x}_2 \pm t^* \sqrt{(\text{s.e.})_1^2 + (\text{s.e.})_2^2}$$ | $H_0 : \mu_1 = \mu_2,$ $$t = \frac{(\bar{x}_1 - \bar{x}_2) - (\mu_1 - \mu_2)}{\sqrt{\frac{s_1^2}{n_1} + \frac{s_2^2}{n_2}}}$$ $$= \frac{(\bar{x}_1 - \bar{x}_2) - (\mu_1 - \mu_2)}{\sqrt{(\text{s.e.})_1^2 + (\text{s.e.})_2^2}}$$ |
| $p$ Proportion | $$(\text{s.e.}) = \sqrt{\frac{\hat{p}(1 - \hat{p})}{n}}$$ | $$p = \hat{p} \pm z^* \times (\text{s.e.})$$ $$= \hat{p} \pm z^* \sqrt{\frac{\hat{p}(1 - \hat{p})}{n}}$$ | $H_0 : p = p_0,$ $$z = \frac{\hat{p} - p_0}{\sqrt{\frac{p_0(1 - p_0)}{n}}}$$ |
| $p_1 - p_2$ Difference of two proportions. | $$(\text{s.e.}) = \sqrt{\frac{\hat{p}_1(1 - \hat{p}_1)}{n_1} + \frac{\hat{p}_2(1 - \hat{p}_2)}{n_2}}$$ $$= \sqrt{(\text{s.e.})_1^2 + (\text{s.e.})_2^2}$$ | $$p_1 - p_2 = (\hat{p}_1 - \hat{p}_2) \pm z^* \times (\text{s.e.})$$ $$= (\hat{p}_1 - \hat{p}_2)$$ $$\pm z^* \sqrt{\frac{\hat{p}_1(1 - \hat{p}_1)}{n_1} + \frac{\hat{p}_2(1 - \hat{p}_2)}{n_2}}$$ | $H_0 : p_1 = p_2,$ $$z = \frac{\hat{p}_1 - \hat{p}_2}{\sqrt{\hat{p}(1 - \hat{p})\left[\frac{1}{n_1} + \frac{1}{n_2}\right]}}$$ where $\hat{p}$ is pooled sample proportion. |

| | | |
|---|---|---|
| The proportion of the population that satisfies $x < A$ | The proportion of the population that satisfies $x > B$ | The proportion of the population that satisfies $A < x < B$ |
| The probability that $x < A$, a.k.a., $$P(x < A)$$ | The probability that $x > B$, a.k.a., $$P(x > A)$$ | The probability that $A < x < B$, a.k.a., $$P(A < x < B)$$ |
| The area under a standard normal curve such that $$z < \frac{A - \mu}{\sigma}$$ | The area under a standard normal curve such that $$z > \frac{B - \mu}{\sigma}$$ | The area under a standard normal curve such that $$\frac{A - \mu}{\sigma} < z < \frac{B - \mu}{\sigma}$$ |
| The probability that $$z < \frac{A - \mu}{\sigma}, \text{ a.k.a}$$ $$P\left(z < \frac{A - \mu}{\sigma}\right)$$ $$P(\mu + z\sigma < A)$$ | The probability that $$z > \frac{B - \mu}{\sigma}, \text{ a.k.a}$$ $$P\left(z > \frac{B - \mu}{\sigma}\right)$$ $$P(\mu + z\sigma > B)$$ | The probability that $$\frac{A - \mu}{\sigma} < z < \frac{B - \mu}{\sigma}, \text{ a.k.a}$$ $$P\left(\frac{A - \mu}{\sigma} < z < \frac{B - \mu}{\sigma}\right)$$ $$P(A < \mu + z\sigma < B)$$ |

# Index

# About the Authors

Bella Bluenose and Romeo Rednose are American Staffordshire Terriers who like mathematics, biscuits, running around in circles for no obvious reason, scratching themselves in odd places, and chasing the occasional squirrel. Romping around campus recently at a major California university, and hearing the pedantic rambles of stats professors through windows and besides dumpsters (as the professors searched for items to pawn for supplemental income, no doubt), they quickly realized that even without vocal cords they could explain things far more clearly than most profs. Much better, in fact, if the reek of fear and indecision coming from inside the classrooms was any indication. Lacking a universal translator (that last ones on Ebay had been purchased by a pod of dolphins due to construction of some sort of interstellar way station), and unable to find a suitable interpreter (those who know APBW, that is, American Pit Bull Woof, are way over-priced), Romeo and Bella decided to write this book. Unfortunately, computer keyboards in their arm of the galaxy are poorly matched to paws and snouts, so expect the occasional typo. And imagine the difficulty a pit bull has getting a decent editor. One bark and they're outta there! Geezy Peezy! So stop complaining about all the typos already and file a report on github if you find one. The form is at https://github.com/bellaromeo/faqsinstats/issues/new. Good grief, they are dogs, what did you expect? And if some of the questions they heard through those windows sounded silly, well, don't blame the messenger. Inquiring student minds wanted to know those sort of things. Hey, was that a squirrel? Woof! Woofwoof!

Romeo Simon Shapiro (2007 - 2017)

www.ingramcontent.com/pod-product-compliance
Lightning Source LLC
Chambersburg PA
CBHW071412180526
45170CB00001B/78